MW00603925

THE TINY FARM Planner

Quarto.com

First Published in 2024 by Cool Springs Press, an imprint of The Quarto Group,
100 Cummings Center, Suite 265-D, Beverly, MA 01915, USA.
T (978) 282-9590 F (978) 283-2742

Cool Springs Press titles are also available at discount for retail, wholesale, promotional, and bulk purchase. For details, contact the Special Sales Manager by email at specialsales@quarto.com or by mail at The Quarto Group, Attn: Special Sales Manager, 100 Cummings Center, Suite 265-D, Beverly, MA 01915, USA.

28 27 26 25 24 1 2 3 4 5

ISBN: 978-0-7603-8901-0

Design: Emily Austin, The Sly Studio
Cover Illustration: Abby Diamond, Instagram.com/abbydiamondart
Page Layout: Emily Austin, The Sly Studio
Illustration: Abby Diamond, Instagram.com/abbydiamondart

Printed in China

THE TINY FARM Planner

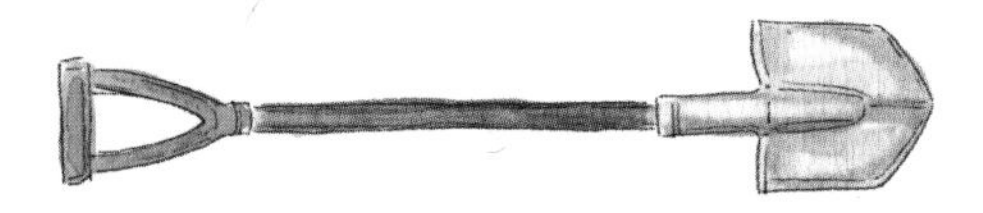

Record Keeping, Seasonal To-Dos, and Resources for Managing Your Small-Scale Home Farm

Jill Ragan

This book belongs to:

Date started:

Contents

Late Season

Winter Season

Postseason

Year-End Thoughts

Things to Improve Next Year

Dream Big: My Farm's Plan for Next Year

Introduction

Hi Friend,

I'm Jill Ragan, farmer and owner of Whispering Willow Farm in central Arkansas's beautiful hill country. For nearly a decade, I've been planting gardens. While my first garden was nothing to write home about, I've honed my skills over the years, developing systems and practices that now yield abundant food and flowers for my family and others year round. I may not have accolades or a fancy résumé, but I was mentored by the best: my grandfather, whom I fondly call "Papaw." From the time I was young, he instilled in me the value of small-scale farming. While it took some years for that lesson to fully sink in, eventually it brought me back to the garden. Today I cultivate more than just produce and flowers—I nurture relationships and build community.

My husband and I farm together full time on just over 4 acres of land. We've spent years making our dreams a reality, and now we want to help you do the same. For me the calling has always been the garden. It's about teaching a generation to be aware of our food sources, being intentional with our surroundings, and ultimately leaving the world better than we found it.

Success didn't come easily; it was earned through hard work, grit, and a level of determination I never knew I needed. My childhood summers were spent at my grandparents', weeding, harvesting, eating, and then starting the cycle all over again. Papaw was meticulous; each crop had its trellis; each row was orderly, and nothing was done in vain. He didn't just throw seeds into the ground; he planned every detail, down to the number of seeds and the required inputs, well before the season started.

Back then, I sometimes wondered why he invested so much time in all this. But once I began farming full time, I understood. The bounty on our kitchen counters and the Christmas gifts of grape jelly were the results of careful preparation that started well before the growing season.

I'm thrilled to have found my way back to the garden. My love for it has rekindled childhood memories that guide me as I cultivate my spaces today. In this planner, I'm going to provide you with all the tools you'll need: charts, logs, spreadsheets, etc., that have helped me transform my humble backyard garden into a profitable farm. And when I say profitable, I don't mean just selling produce at the farmers' market or having numerous wholesale accounts. We don't even need to go to the grocery store for food; we simply walk out our back door. I hope to inspire and encourage you to pursue this fulfilling lifestyle as well. With the right systems and a little guidance from your garden bestie (that's me!), you too can start growing your own food right in your backyard.

Here's to making spaces abundant!

Jill

Quick Inspiration and Goals

It's time to dream big and let your imagination flourish. What does your dream garden look like? What memories will it hold? Jot down your goals for today on this page. Then, let's break them into achievable milestones for next year by writing future goals on the next page. From there, turn the page to the next spread where you'll track your longer-term goals for 5 and 10 years from now. Let's get started!

Goals You Can Reach Today

- ☐ ....
- ☐ ....
- ☐ ....
- ☐ ....
- ☐ ....
- ☐ ....
- ☐ ....
- ☐ ....
- ☐ ....
- ☐ ....
- ☐ ....

Goals You Can Reach in 1 Year

- ☐ ....
....
- ☐ ....
....
- ☐ ....
....
- ☐ ....
....
- ☐ ....
....
- ☐ ....
....
- ☐ ....
....
- ☐ ....
....
- ☐ ....
....
- ☐ ....
....
- ☐ ....
....
- ☐ ....
....
- ☐ ....
....
- ☐ ....
....

Goals You Can Reach in 5 Years

- ☐
- ☐
- ☐
- ☐
- ☐
- ☐
- ☐
- ☐
- ☐
- ☐
- ☐
- ☐
- ☐
- ☐

Goals You Can Reach in 10 Years

- ☐
- ☐
- ☐
- ☐
- ☐
- ☐
- ☐
- ☐
- ☐
- ☐
- ☐
- ☐
- ☐
- ☐

12-Month at-a-Glance Calendar

JANUARY						
1	2	3	4	5	6	7
8	9	10	11	12	13	14
15	16	17	18	19	20	21
22	23	24	25	26	27	28
29	30	31				

FEBRUARY						
1	2	3	4	5	6	7
8	9	10	11	12	13	14
15	16	17	18	19	20	21
22	23	24	25	26	27	28
29						

MAY						
1	2	3	4	5	6	7
8	9	10	11	12	13	14
15	16	17	18	19	20	21
22	23	24	25	26	27	28
29	30	31				

JUNE						
1	2	3	4	5	6	7
8	9	10	11	12	13	14
15	16	17	18	19	20	21
22	23	24	25	26	27	28
29	30					

SEPTEMBER						
1	2	3	4	5	6	7
8	9	10	11	12	13	14
15	16	17	18	19	20	21
22	23	24	25	26	27	28
29	30					

OCTOBER						
1	2	3	4	5	6	7
8	9	10	11	12	13	14
15	16	17	18	19	20	21
22	23	24	25	26	27	28
29	30	31				

MARCH						
1	2	3	4	5	6	7
8	9	10	11	12	13	14
15	16	17	18	19	20	21
22	23	24	25	26	27	28
29	30	31				

APRIL						
1	2	3	4	5	6	7
8	9	10	11	12	13	14
15	16	17	18	19	20	21
22	23	24	25	26	27	28
29	30					

JULY						
1	2	3	4	5	6	7
8	9	10	11	12	13	14
15	16	17	18	19	20	21
22	23	24	25	26	27	28
29	30	31				

AUGUST						
1	2	3	4	5	6	7
8	9	10	11	12	13	14
15	16	17	18	19	20	21
22	23	24	25	26	27	28
29	30	31				

NOVEMBER						
1	2	3	4	5	6	7
8	9	10	11	12	13	14
15	16	17	18	19	20	21
22	23	24	25	26	27	28
29	30					

DECEMBER						
1	2	3	4	5	6	7
8	9	10	11	12	13	14
15	16	17	18	19	20	21
22	23	24	25	26	27	28
29	30	31				

Preseason

Welcome to the preseason phase of your garden planner, where excitement meets groundwork in the journey toward a thriving garden. As the earth shakes off the winter chill, gardeners eagerly prepare to reconnect with their outdoor spaces. Preseason marks the crucial moment when plans take shape, seeds are chosen, and the canvas of your garden eagerly awaits its transformation.

During this transitional period, the focus shifts from indoor planning to outdoor planting. It's a time for assessing the aftermath of winter, clearing away debris, and revitalizing the soil for the upcoming growing season. Preseason awaits the start of a journey where gardeners become stewards of their own land, embracing the ebb and flow of growth and renewal.

Amidst the flurry of preparation, there's a tangible sense of anticipation. Gardeners pore over seed catalogs, envisioning the kaleidoscope of colors and textures that will soon adorn their beds. They meticulously plan layouts, considering the needs and interactions of each plant and envisioning the vibrant possibility of life that will soon grace their garden.

However, preseason isn't just about physical tasks; it's also a time for reflection and intention setting. It's an opportunity to reconnect with the rhythms of nature, to align our actions with the cycles of the seasons, and to deepen our understanding of the interconnectedness between gardener and garden.

So, as you embark on this journey through the preseason, embrace the anticipation, relish the groundwork, and let the promise of new growth ignite your passion for gardening.

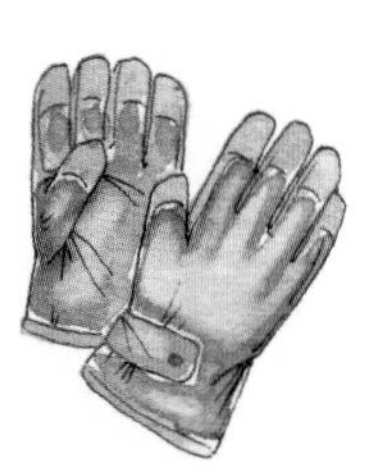

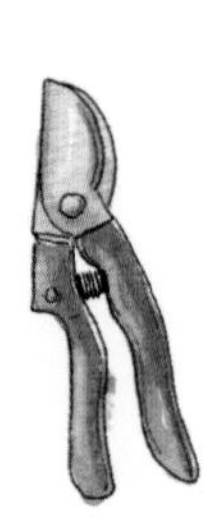
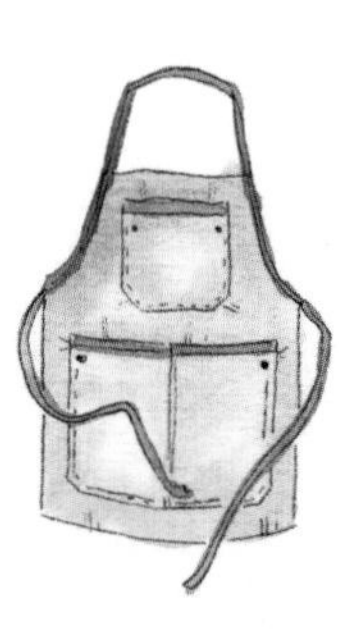

"Before the first seed
touches the soil, the
planner paints the
canvas of possibility,
where every stroke
is a thoughtful
intention for a
fruitful harvest."
—Anonymous

Farm To-Dos for Preseason (Waking Up Your Farm)

In the preseason, your garden vision is paramount. You've spent months rummaging through seed catalogs, gathering inspiration, and sketching out plans. As the early season approaches, you have the chance to change with the changing seasons. Now you can take all that vision and planning and take action; you can finally plant the seeds of your imagination to cultivate a bountiful garden.

- ☐
- ☐
- ☐
- ☐
- ☐
- ☐
- ☐
- ☐
- ☐
- ☐
- ☐

- ☐
- ☐
- ☐
- ☐
- ☐
- ☐
- ☐
- ☐
- ☐
- ☐
- ☐
- ☐
- ☐
- ☐
- ☐

How to Test Your Soil

I'm passionate about soil, perhaps more than one would normally be about the earthy stuff beneath our feet. But the fact is, everything begins in the soil. It's a vital component whose health should never be overlooked. Soil should be the top priority on any farm or in any garden. Whether you dream of a bountiful harvest, or rows of vibrant cut flowers, or you simply find solace tending your garden, it all starts with the quality of your soil.

One reason I place such emphasis on soil health is that it is the backbone of your farm or garden and it's a long-term investment. It takes months or even years of adding organic matter and nutrients, and employing practices like cover cropping, amending, and consistent fertilization to truly get your soil up to snuff. So it's critical to start now, by testing, treating, and emphasizing the health of your soil.

Before treating your soil, it's crucial to understand its current state. Just like you wouldn't begin treating an illness without identifying the symptoms, you need to know what nutrients your soil has in abundance and which it lacks. A soil test provides this essential base knowledge, allowing you to tailor your soil treatments accordingly.

Testing your soil is simple, and you'll likely find most needed supplies in your pantry.

First, gather a plastic bag (or two), a good-quality marker, and a garden trowel, for collecting soil samples.

Quick Tip:

Make sure you collect enough soil for a meaningful test. I usually collect 1 to 2 cups of soil from each area I'm testing.

Now it's time for the fun part: getting your hands dirty. Collect soil samples from various locations in your garden. I like to collect separate samples from my north, south, east, and west beds. I highly recommend testing the soil from every area where you plan to plant. Dig soil from the root zone, at a depth of 4 to 6 inches (10 to 15 cm). Don't forget to label each bag with the corresponding area of your garden!

You can either test your samples at home using a soil test kit or send them off to a lab for more accurate results. Laboratories charge a fee, but give you far more comprehensive data. The lab can also tailor your results to what you want to grow.

When submitting your soil samples, consider specifying that you'd like to test for nutrients like nitrogen, phosphorus, and potassium (N-P-K), which your plants utilize for various purposes. Also it's crucial to check the pH level of your soil; this factor critically influences your plants' nutrient availability.

Soil Test Results Chart

SOIL TEST DATE	GARDEN SECTION	pH ADJUSTMENTS	SOIL TYPE	pH LEVEL	NUTRIENT LEVELS	AMENDMENTS NEEDED	ORGANIC MATTER NEEDED

Soil Amendment Types

When it comes to soil amendments, navigating the multitude of options, organic or inorganic, can be a daunting task. It's challenging to distinguish the good from the bad. So let's simplify this. Here I'll share a few of my favorite types of soil amendments, what they are, what they do, and how they affect your soil.

Feather Meal: Making feather meal involves partially grinding poultry feathers under heat and pressure, followed by drying. It acts as a slow-release fertilizer, enriching the soil with nitrogen.

Bone Meal: This meal comprises finely and coarsely ground animal bones that have been steamed, dried, and pulverized. The impact on your soil is significant. Bone meal is an excellent source of calcium and phosphorus and is slow to release its nutrients.

Blood Meal: Made from animal blood, often livestock blood, blood meal is a product that would otherwise go to waste. It's a potent organic fertilizer, rich in nitrogen, and offers slow-release benefits.

Kelp Meal: Made from dried and ground seaweed, kelp meal is a fertilizer that's hard to overuse, making it ideal for beginners. It's loaded with more than 60 trace minerals or micronutrients essential for plant health.

Compost: Made of decomposed organic materials, compost is full of micro- and macronutrients. It feeds your soil gradually as a slow-release amendment.

Wood Ash: Boasting phosphorus, calcium, magnesium, potassium, and various essential nutrients, wood ash makes an excellent soil amendment. Use caution to not overapply or you might damage your soil's pH.

Azomite®: Azomite® is a broad-spectrum trace mineral product. Trace minerals are vital for robust plant growth.

Rice Hulls: A byproduct of rice milling, rice hulls are sterilized with steam and dried. Much like perlite, they enhance aeration and drainage, critical in heavy clay soils.

Worm Castings: Worm castings are an inexpensive and long-term soil amendment. You can get worm castings by setting up a worm farm. They improve soil structure, aeration, and nutrient levels, offering a wealth of benefits to your soil.

These are just a few among many organic amendment options. Keep exploring to find what suits your soil type and specific needs. I always advocate for incorporating organic amendments to enrich and nurture your soil. Just as there are numerous organic amendment options, the same holds true for synthetic ones. Be mindful of how these amendments affect your soil quality and the natural habitat around it. Balance and consideration are key to cultivating a healthy garden ecosystem.

Soil Amendment Tracking Chart

DATE	AMENDMENT APPLIED	AMOUNT APPLIED	LOCATION	NOTES

Trellis System Options for the Farm

In small-scale farming and gardening, making the most of available space is a constant challenge. One highly effective technique to optimize small spaces and maximize yield is trellising. Trellising provides a framework or support for plants to grow vertically, therefore utilizing the vertical space more efficiently. Not only does this maximize space, but it also improves airflow and exposure to sunlight and facilitates easier maintenance and harvesting.

Lower and Lean

One popular trellising system, known as the "lower and lean" method, is ideal for growing tall and vining plants. In this system, you prune plants to a single branch, and train them up a piece of twine, string, or guide wire. As the plant grows heavy and starts reaching the top of the string, you release additional string from a spool and reattach it along the trellis guidewire. Then you lower the plant and lean it over. This method allows for better management of plant growth and creates a more productive growing environment.

Ideal Plants for the Lower and Lean Method

Tomatoes

Cucumbers

Peppers (vining varieties)

Two-Leader

Another efficient trellising system is the "two-leader" method, great for tomato growing. Before you transplant seedlings, prune their tops. This encourages the plant to develop two main leaders growing upward. To do this, attach two lengths of twine at the base of the plant, extending to the tops of the conduit or pipe supports. Train both leaders upwards as they grow, using tomato clips. This method provides excellent support and encourages optimal growth.

Ideal Plants for the Two-Leader System

Tomatoes

Cattle Panel Trellises

Cattle panel trellises are popular for their aesthetic appeal. These panels, typically made of welded galvanized wire, can be easily found at local hardware or feed stores. They often span 16 feet (5 m), forming an arch over raised or in-ground beds. This type of trellis offers excellent support and structure for a variety of climbing plants.

Ideal Plants for Cattle Panels

Cucumbers	Beans	Melons
Peas	Squash	Vining Flowers

Florida Weave

If you want to grow determinate tomatoes in a compact space, the “Florida weave” trellis system, also known as the basket weave method, is ideal. This technique involves placing stakes or T-posts at row ends and between every two tomato plants, and then weaving twine in a looping pattern to create a sturdy support structure.

Ideal Plants for the Florida Weave

Tomatoes

Net Trellis

Trellis netting is a cost-effective and versatile option for growing plants vertically. Its lightweight polypropylene mesh provides ample support for climbing vines and vegetables, making it an excellent choice to maximize space and improve accessibility for maintenance and harvesting.

Ideal Plants for Trellis Netting

Cucumbers	Peas	Tomatoes
Beans		

Umbrella Trellising

This method is one of the most appealing and efficient ways to cultivate cucumbers. Not only is it visually pleasing, it also simplifies harvesting and maintenance. Much like the two-leader system, this approach involves setting up conduit or pipe supported by PVC T-joints placed on top of T-posts that are firmly staked into your raised bed or in-ground garden. To use this system, you affix a string from the top to the bottom of the conduit and secure it using a landscape staple. As the cucumber plant begins to grow, fasten it to the twine using a clip. When the plant reaches the top of the pipe or conduit, provide a gentle nudge to guide it over, and the plant will cascade down in an “umbrella” fashion.

Trellising systems offer a range of options to optimize space and improve yield in a farm or garden setting. From the “lower and lean” method to cattle panels, the choice of trellising system largely depends on the type of crops being grown and the available space.

Trellising System Log

DATE	TRELLIS TYPE	SUPPLIER	TASK	NOTES

DATE	TRELLIS TYPE	SUPPLIER	TASK	NOTES

Seed Inventory

Seasoned or newbie, every gardener knows that planning is the heartbeat of a successful garden. And within that plan, tallying what you have in stock and what you need for the upcoming season is crucial. I can't stress enough how vital it is to get a head start on this. Forgetting to order seeds can lead to disappointment, with your desired crops out of stock or struggling to find the right varieties.

I've been through this and, trust me, you want to avoid this. Over time I've learned the importance of staying organized, especially starting with seeds and inputs, a principle that carries through each season. That's why this section is so important—it's about helping you keep a close eye on what seeds you have, their unique characteristics, and when to plant them. It's like having a gardening best friend to help keep you on track.

Having a detailed inventory list is extremely helpful. It means you never second-guess what you have or where to find it, bringing a sense of order to your gardening adventure. This knowledge empowers you to spend wisely, plan your garden effectively, and make thoughtful choices, like the best times for planting and how to create harmonious companion plantings, and more.

Seed Inventory Chart

CROP	VARIETY	SUPPLIER	QUANTITY	COST	PURCHASE DATE	NOTES

Seed Inventory Chart

CROP	VARIETY	SUPPLIER	QUANTITY	COST	PURCHASE DATE	NOTES

Seed Inventory Chart

CROP	VARIETY	SUPPLIER	QUANTITY	COST	PURCHASE DATE	NOTES

Seed Inventory Chart

CROP	VARIETY	SUPPLIER	QUANTITY	COST	PURCHASE DATE	NOTES

Seed Inventory Chart

CROP	VARIETY	SUPPLIER	QUANTITY	COST	PURCHASE DATE	NOTES

Tool Inventory

This section is all about practicality. While having the latest tools is nice, gardening success doesn't hinge on them. Find those essential truly efficient tools. Evaluate, prioritize, and personalize your toolkit for a productive gardening experience. For me, daily essentials include my trusty toolbelt, sharp pruners, a reliable permanent marker, and a sturdy knife.

Tool Inventory Chart

TOOL	QUANTITY	SUPPLIER	CONDITION	MAINTENANCE RECORD

Tool Inventory Chart

TOOL	QUANTITY	SUPPLIER	CONDITION	MAINTENANCE RECORD

Preseason Planning

As you embark on your tiny farm journey, it's vital to set clear and inspiring goals for the season ahead. Think about what you want to achieve with your garden. Do you desire a wealth of fresh, homegrown produce for your family's table? Maybe you're eager to preserve the harvest by canning and fermentation, or perhaps you dream of sharing the fruits of your labor with your community, selling your produce at the local farmers' market. Whatever your aspirations, define them with clarity and purpose. Be specific about your intentions. Writing down your goals will transform them from aspirations into guiding directions and actionable steps.

With your goals in focus, it's time to craft a planting strategy that aligns with your vision. Select crops that complement your goals and thrive in your climate. If canning is your aim, opt for high-yield, hybrid varieties of tomatoes or cucumbers. If you crave the simple pleasure of fresh eating, explore heirloom varieties with countless flavors and variety selections. Ensure your garden becomes a feast for the senses, delighting both your taste buds and your soul.

As you chart your garden journey, embrace the beauty of diversity. Mix in herbs, flowers, and companion plants to foster biodiversity, attract pollinators, and naturally deter pests. Create an ecosystem teeming with life, and it will not only nurture your primary crops but create a vibrant and visually stunning sanctuary in your backyard. Let your garden be a tapestry of harmonious relationships where plants support and nurture each other, creating a resilient and thriving microcosm. With each decision, remember that you're not merely planting seeds, you're cultivating connections and weaving a story of growth and abundance.

Crafting Your Garden Plan

Here are some quick tips to set yourself and your gardens up for success this season and beyond.

- **Know Your Space:** Assess your garden area thoroughly. It's crucial to provide each crop with optimal sunlight and good drainage, preventing any unwanted flooding. Understanding the unique requirements of the crops you plan to grow allows you to tailor your garden space to meet their needs, ensuring a bountiful harvest.
- **Planting Times:** Research your growing zone and the best times to plant each crop in your region. This ensures a steady supply of produce throughout the season.
- **Crop Rotation:** If you plan to grow the same crops next year, practice crop rotation to improve soil health and reduce pest problems.
- **Succession Planting:** For a continuous harvest, stagger planting times of the same crop at regular intervals.
- **Keep Learning:** Stay curious and learn from fellow gardeners, improving your knowledge with each passing season.

Preseason Planning Notes

A Visual Plan of the Garden

Use this spread to sketch out your garden's design. Remember to draw in pencil so you can easily alter your plan if necessary.

Tips on How Much to Plant for Your Family

Planning the right amount of produce for your family can be a delightful challenge, regardless of your gardening experience. Factors such as preservation plans, consumption preferences, and even the willingness of your children to try new vegetables come into play. Throughout this planner, I'll emphasize the critical role of planning. Successful gardening often hinges on well-thought-out plans made before the growing season kicks in.

One crucial piece of advice for determining how much to plant for your family is to understand your family's dietary preferences. Identify the staples your family frequently consumes from the grocery store or farmers' market. Consider your overarching goals for this growing season. Understanding what you intend to preserve is equally important as it helps shape your seasonal plan.

For a family of four, a garden area of around 800 square feet (74 sq m) is a good starting point, translating to about 200 square feet (19 sq m) of garden space per person. However, this can vary based on such factors as weather, soil quality, pest control, chosen crop varieties, and more. Keep these variables in mind while planning your crop quantities.

If this is your initial attempt at maximizing your homegrown produce, give yourself some grace. Getting the quantities right in the first season can be challenging, and even seasoned gardeners miscalculate. The key is to persist and fine-tune your approach. Identify the varieties that thrive in your region, and over time, you'll achieve a sustainable garden that provides for your family year-round.

To help you in your planning, I've calculated the approximate number of plants needed for a family of four for some common crops. However, these are just estimates, and you should adjust them according to your family's needs and preferences.

To calculate the number of plants for any variety you intend to grow, follow these steps:

- ☐ Determine the weekly consumption per person of the specific crop for your family.
- ☐ Multiply the weekly consumption per person by the number of family members.
- ☐ Multiply this by the number of weeks in the growing season.
- ☐ Research and find the estimated yield per plant for the chosen crop variety.
- ☐ Divide the total annual consumption by the average yield per plant to estimate the number of plants needed.

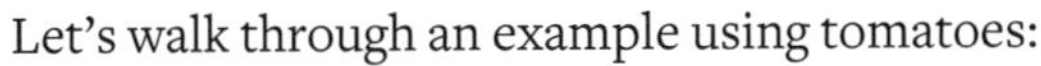

Let's walk through an example using tomatoes:

A healthy tomato plant can yield about 10 to 20 pounds (4.5 to 9 kg) of tomatoes, and hybrid varieties can yield even more. Assuming each person consumes about 1 pound (454 g) of tomatoes per week, we can calculate the number of plants needed for a family of four for a year:

- ☐ Total tomatoes needed per year: 4 (people) × 1 (pound [454 g] per week) × 52 (weeks) = 208 pounds (94 kg)
- ☐ Tomato plants needed: 208 pounds (94 kg) / 15 pounds (7 kg) per plant (average) ≈ 14 plants (rounded up for better yield)

Remember, you can adjust these calculations based on your specific circumstances and goals. Gardening is an evolving journey and finding the right balance is part of the joy of cultivating your own produce.

Quantities for a Typical Family of Four

VEGETABLE	PLANTS NEEDED (ESTIMATED)
Tomatoes	14
Cucumbers	14
Peppers	17
Potatoes	30
Onions/Garlic	104
Squash	14

Seasonal Transition: Preseason to Early Season

In the preseason, casting a vision is paramount. You've spent months indoors, rummaging through seed catalogs, gathering inspiration, and sketching out plans. But as the early-season approaches, you get to move from vision to action. Now it's time to take all that vision and planning and turn it in to action; you can finally plant the seeds of your imagination to cultivate a bountiful garden.

As the first daffodils and tulips emerge, they mark a long-awaited transition: from winter's gloom to spring's renewing light, each seed sown carrying a renewed sense of purpose. It's a moment to pause and contemplate the promise of the season ahead.

Here comes my favorite part: the chance to learn and truly grow as a gardener. The early season is all about navigating priorities. In my book *The Tiny but Mighty Farm*, I describe soil as the "life and longevity" of your farm. My sentiments on this haven't changed; soil health should be at the forefront of your mind in the early season. Once you've sent off your soil tests and received the results, it's time to amend your beds with organic matter, compost, and organic fertilizers. Your role as the gardener is to create the most welcoming environment for your plants to thrive. If you're using raised beds, now is the time to top-dress them with new soil and nutrients, preparing the way for your soon-to-be plant babies.

Next comes the planting phase, where all your planning truly starts to pay off. By now, you should know your crop plan like the back of your hand, understanding where each plant will go. This is your chance to put all that preliminary planning to the test. One thing I've learned over the years is that when the garden wants to teach you something, you'd better listen. While it's important to know your crop plan, garden layout, and optimal sunlight conditions, you should also remain somewhat flexible. Gardens often require adaptability. Have a plan, but if the need to pivot arises, welcome it with open arms.

You've dreamed, planned, amended soil, and planted seeds. The next step in the early season is to ensure proper watering for your plants. By this stage, you should already have an irrigation plan in place. Drip hoses on automatic timers are my personal favorites, but there are also soaker hoses and traditional watering methods to consider. Being realistic about the demands and size of your garden is crucial. This is one area where investing a bit more can pay off significantly, resulting in an abundance of produce.

The idea of the early season is to put your preseason plans to the test. Observe and listen: What is your garden telling you? Are your seedlings sprouting with health and abundance, or are they diseased and under attack from pests? Is your garden receiving enough water, or should you consider automating your system? Now is the time to adjust and revise before the hustle and bustle of mid-season arrives. If you're new to gardening, take a deep breath. You're not going to get everything right the first time, and that's okay. As you grow, so will your garden.

Month-by-Month Early-Season Checklist

Jill's Checklist

Month 1: Planning and Preparation

- ☐ Review and adjust crop rotation plan.
- ☐ Order seeds and supplies for the upcoming growing season.
- ☐ Prepare an indoor seed-starting area or greenhouse with soil, trays, heat mats, lights, etc.
- ☐ Prune fruit trees and dormant plants.
- ☐ Repair and maintain farm equipment.
- ☐ Plan irrigation systems and make any necessary repairs.
- ☐ Create a schedule for composting and soil amendments (when to turn, when to add amendments seasonally, etc.)
- ☐ Review and update your farm budget for the year.

Month 2: Starting Seeds and Soil Prep

- ☐ Sanitize and clean seed-starting supplies.
- ☐ Start indoor seeds for early spring crops.
- ☐ Collect a soil test and amend the soil as needed.
- ☐ Begin greenhouse or cold frame preparation.
- ☐ Prune berry bushes and grapevines.
- ☐ Plan your layout and design new garden beds or expansions.
- ☐ Organize tools and inventory.
- ☐ Attend gardening workshops, conferences, and online webinars and gain knowledge for the upcoming season.

Month 3: Early Planting and Garden Cleanup

- ☐ Transplant seedlings to outdoor beds.
- ☐ Direct sow hardy crops like peas and lettuce.
- ☐ Clear garden beds of debris and weeds.
- ☐ Prune and shape shrubs and ornamental plants.
- ☐ Start compost turning and monitor decomposition.
- ☐ Check for signs of early pests and disease; apply treatments if necessary.
- ☐ Begin training and pruning fruiting plants.
- ☐ Evaluate preservation for the upcoming season and buy supplies.

My Checklist

☐ ....

☐ ....

☐ ....

☐ ....

☐ ....

☐ ....

☐ ....

☐ ....

☐ ....

☐ ....

☐ ....

☐ ....

☐ ....

☐ ....

☐ ....

☐ ....

☐ ....

☐ ....

☐ ....

☐ ....

☐ ....

☐ ....

☐ ....

☐ ....

☐ ....

☐ ....

☐ ....

☐ ....

Early-Season Friendly Reminders

Test and amend your soil for spring plantings. The early season is the perfect time for this.

Start those longer germinating seeds indoors, like tomatoes, peppers, and eggplants.

Prune your fruit trees, berry bushes, and roses. Remove any dead branches to encourage new growth.

Grab that pen and notebook. Now's the time to start planning your garden layout and make any needed changes from the previous year.

Don't forget to consider crop rotation, companion planting, and how best to utilize your small space.

Don't forget about those unexpected late frosts; keep your row covers handy.

Order any seed-starting supplies you may need early, depending on when you plan to start your seeds. Things to have on hand would be seedling heat mats, grow lights (if applicable), soil blocks, pots and trays, labels, pens, etc.

Start sowing those cold-weather vegetables like spinach, lettuce, and kales for an early spring harvest.

Start creating your planting calendar. Note when to start seeds, when to transplant them, and when to add succession sowings.

Don't forget about flowers. They attract pollinators and add a depth of beauty to any garden. Begin sowing flower seeds indoors.

Take inventory of your garden tools. Ensure they are cleaned, sharpened, and organized.

Create a harvest plan. Begin by developing a strategy based on efficient harvesting and how you intend to preserve the harvest. This is important when you are thinking through your succession plantings as well to give yourself enough time to harvest and preserve before your next harvest is ready.

Clean and sanitize the pots and trays you plan on using for seed starting. This will help prevent the spread of disease in your young seedlings.

Check and clean your greenhouse or indoor growing spaces to ensure they are ready for the start of the season.

To sanitize trays, gather a black tote, and fill with water for pre-rinsing. In another tote add 1 quart (1 L) hydrogen peroxide, 1 cup (236 ml) disinfectant concentrate, and 2 gallons (7.5 L) of water. Add pots and trays to the water first to remove dirt and grime, then put them into the solution mix to disinfect. Air dry.

Be prepared for a late frost. Have your row covers on hand if you need to cover any crops.

Start composting your kitchen scraps, garden waste, and other organic materials to create nutrient-rich compost for your garden beds later in the season.

Buy or create your seed-starting soil. I like to buy our soil in bulk in the early season to save money and avoid running out.

Spend time learning about succession planting for both vegetables and flowers so you can have a continuous harvest throughout the season.

Check any bulbs you stored for the winter. Discard any damaged or rotting bulbs.

Check irrigation lines for any leaks or holes and replace as needed before the busy season begins.

Research and jot down your average last frost dates in your zone to plan your outdoor plantings accordingly.

Research and order any organic fertilizers you may need later in the season.

Determine the types of supports and trellising you'll need for the season and make sure you have all the supplies ordered or on hand.

Keep dreaming for the next season! Although you are in the early-season hustle and bustle, the mid-season is right around the corner. Set realistic goals for yourself. Ultimately give yourself grace, embrace the victories of this season, and stay encouraged—there is more to come!

Chore Tracking

I find that organizing my chores by the day of the week is helpful for staying on track. Maybe Wednesday is weeding day and Tuesday is harvesting day on your farm. Use this chart to track your early-season chores and remember what to do on each day of the week. It will be helpful as the season progresses to know what needs to be done when. Chores in the Any Day column can be done whenever you have a bit of extra time in your schedule.

MONDAY	TUESDAY	WEDNESDAY	THURSDAY

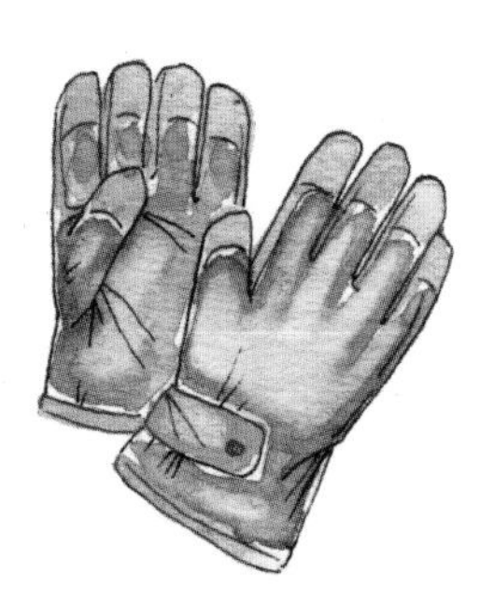   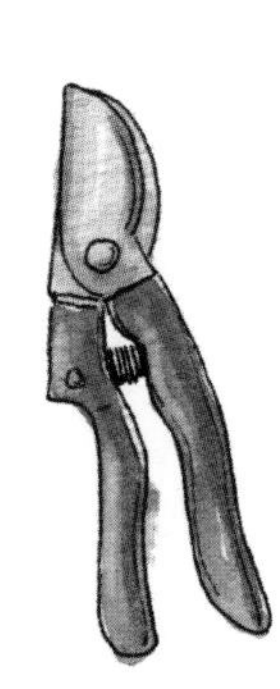 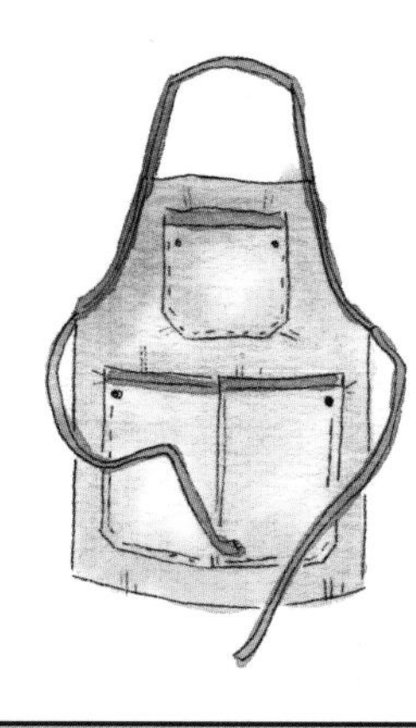

FRIDAY	SATURDAY	SUNDAY	ANY DAY

My Farm's Early-Season Inventory: What We Have

SEEDS	SOIL	SOIL AMENDMENTS	POTS &

   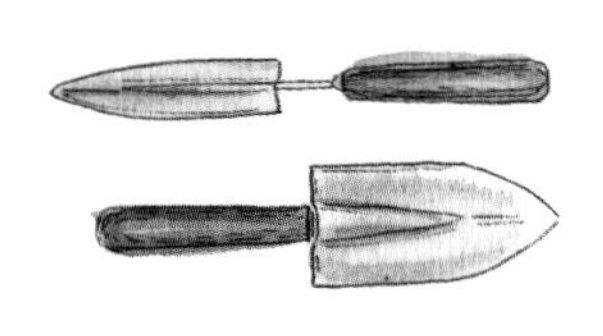  

CONTAINERS	TOOLS	IRRIGATION SUPPLIES	ADDITIONAL ITEMS

Early-Season Supply Needs: What We Need

SEEDS	SOIL	SOIL AMENDMENTS	POTS &

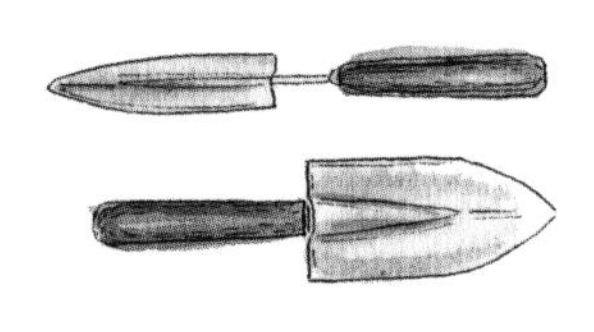

CONTAINERS	TOOLS	IRRIGATION SUPPLIES	ADDITIONAL ITEMS

Early-Season Expense Log

DATE	CATEGORY	VENDOR	ITEM PURCHASED

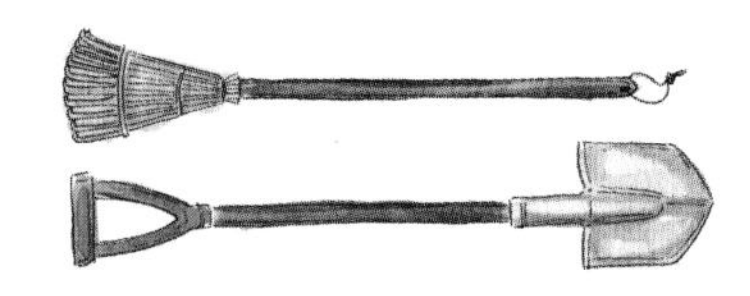

QUANTITY	TOTAL COST	PAYMENT METHOD	NOTES

Seed-Starting Schedule

Use this calendar to record your seeding dates, numbers of starts, how you planted and location in which you sowed them out. It's important to use pencil as these change as you lay out your gardens. You may need to erase and rewrite a few times until your final planting calendar is final.

TP = Transplant   DS = Direct Seed

CROP	SEEDING DATE	SIZE TRAYS	# OF STARTS	TP OR DS	LOCATION	SUCCESSION PLANTING

CROP	SEEDING DATE	SIZE TRAYS	# OF STARTS	TP OR DS	LOCATION	SUCCESSION PLANTING

Companion Planting Chart

This companion planting chart is a tool to guide you in strategically pairing plants for mutual benefit while avoiding harmful combinations. Utilize the chart to refine your interplanting approach, promoting pest management, disease prevention, and optimized growth. Boost your garden to be healthier and more fruitful.

CROP	COMPANION PLANTS	INCOMPATIBLE PLANTS	PLANTING DATE	BENEFITS

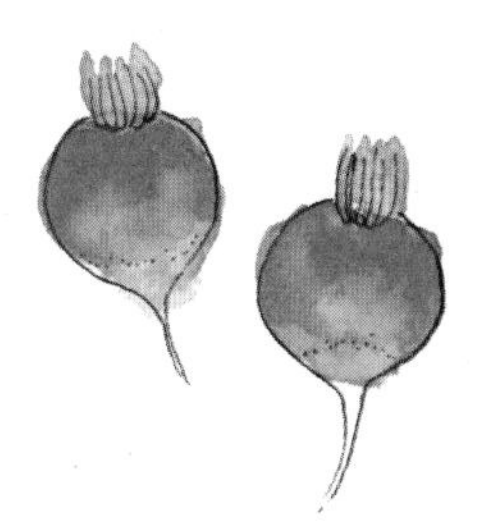

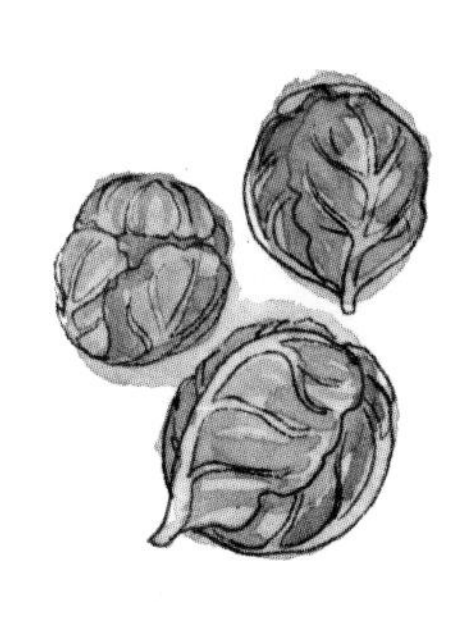

CROP	COMPANION PLANTS	INCOMPATIBLE PLANTS	PLANTING DATE	BENEFITS

Early-Season Pest and Disease Tracker

DATE	PEST/DISEASE	PLANT AFFECTED	SYMPTOMS

NTENSITY LEVEL	TREATMENT	PREVENTION	NOTES

Early-Season Variety Reviews

CROP	VARIETY	LOCATION	PESTS	DISEASES	FLAVOR

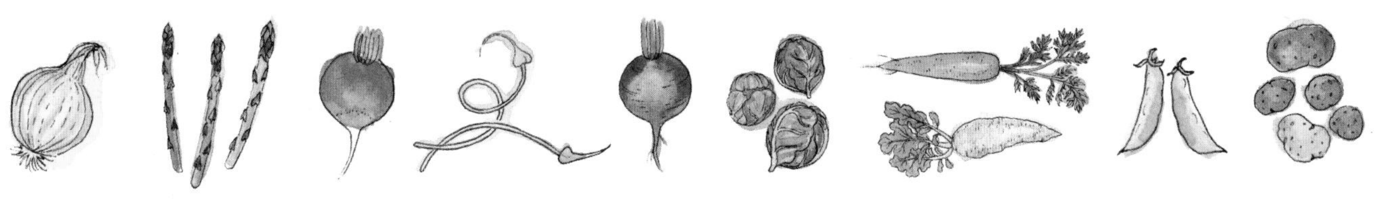

PERFORMANCE	YIELDS	PROS/CONS	EXTRA NOTES

Early-Season Variety Reviews

CROP	VARIETY	LOCATION	PESTS	DISEASES	FLAVOR

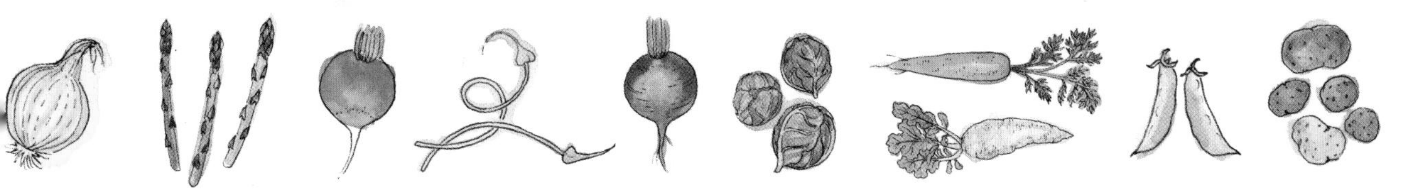

PERFORMANCE	YIELDS	PROS/CONS	EXTRA NOTES

Planting Calendar Chart

CROP	INDOOR START DATE	TRANSPLANT DATE	DIRECT SEED DATE	SPACING

IL REQUIREMENTS	SUN REQUIREMENTS	WATER REQUIREMENTS	NOTES

Planting Calendar Chart

CROP	INDOOR START DATE	TRANSPLANT DATE	DIRECT SEED DATE	SPACING

IL REQUIREMENTS	SUN REQUIREMENTS	WATER REQUIREMENTS	NOTES

Seasonal Transition: Early Season to Mid-Season

> "The love of gardening is a seed once sown that never dies."
> —Gertrude Jekyll

Well friends, we're officially in the peak season now. The early days of harvesting and weeding have become a daily routine. We're right in the middle of it, as they say. The excitement rushes through your body as you harvest the first ripe cucumber from the vine. However, as the months go by, that excitement can quickly turn into feeling overwhelmed and exhausted.

This is the time to keep a steady pace and remember that gardening is a marathon, not a sprint. Each day, ferment, freeze, or preserve the produce you can. Focus on pruning and weeding regularly to stay on top of those tasks. Take it one day at a time, working on streamlining your garden efforts to make this peak season enjoyable. I've spent years feeling overwhelmed by the never-ending task list of the mid-season, but once I became efficient and created daily chore charts, it made a significant difference. I can't stress enough how important it is to prioritize your time wisely.

Yes, it's a whirlwind, but remember, those sun-kissed tomatoes and beautiful dahlias won't be around forever. Let's make the most of it! Here's a glimpse into my daily tasks during the peak season. My priorities shift as we move throughout the seasons, but I hope this gives you a head start in creating your own systems for a successful garden experience.

Monday: I spend most Mondays in the greenhouse. Throughout this season, I do weekly and biweekly successions of lettuce crops and flowers like sunflowers and zinnias. This is the day to get these crops seeded and transplant the prior week's starts into the garden.

Tuesday: Mornings start early, harvesting flowers before the Sun comes up, and I make my twice-a-week delivery of wholesale orders.

Wednesday: This day is affectionately named "Weeding Wednesdays" on the farm. It's a day to cultivate, prep beds if needed, and check drip lines for any issues.

Thursday: Like Tuesdays, Thursdays are harvest and delivery days during this time of year. Flowers are harvested early in the morning and head out to wholesale markets. We also have a few customers who pick up on the farm later in the afternoon.

Friday: This is a critical day because we can evaluate all that we've accomplished and note what needs to be added to next week's tasks. We'll walk the farm with a notebook in hand, reassess priorities, and make a game plan for the following week.

Month-by-Month Mid-Season Checklist

Jill's Checklist

Month 1: Spring Planting and Maintenance

- ☐ Plant cool-season crops and warm-season transplants.
- ☐ Install drip irrigation or soaker hoses.
- ☐ Mulch garden beds to conserve moisture and suppress weeds.
- ☐ Monitor and manage pests.
- ☐ Begin harvesting early spring crops.
- ☐ Evaluate preserving supplies and buy as needed.
- ☐ Implement a mowing and weeding maintenance schedule.
- ☐ Connect with local farmers and gardeners.

Month 2: Growth and Expansion

- ☐ Plant heat-loving crops like tomatoes and peppers.
- ☐ Thin seedlings and space plants properly.
- ☐ Train vines and trellis-climbing plants.
- ☐ Fertilize and feed plants as needed.
- ☐ Harvest and process early crops.
- ☐ Plan and set up summer cover crops.
- ☐ Continue preserving your early-season crops.
- ☐ Participate in local farmers' markets or community events.

Month 3: Care and Harvesting

- ☐ Monitor and manage irrigation during hotter months. Water more frequently if needed.
- ☐ Continue planting successions of crops for continuous harvest.
- ☐ Prune and train fruit trees and berries for proper growth.
- ☐ Harvest and process summer crops regularly.
- ☐ Assess for and address any disease or pest issues.
- ☐ Develop a plan for fall crops and seed starting.
- ☐ Review and update farm sustainability practices you want to improve on.
- ☐ Add shade cloth to high tunnels and greenhouses.

My Checklist

- ☐
- ☐
- ☐
- ☐
- ☐
- ☐
- ☐
- ☐
- ☐
- ☐
- ☐
- ☐
- ☐
- ☐
- ☐
- ☐
- ☐
- ☐
- ☐
- ☐
- ☐
- ☐
- ☐
- ☐
- ☐
- ☐
- ☐
- ☐

Mid-Season Friendly Reminders

Anytime is a good time to top-dress beds with quality compost and amendments.

Gardening is just as much about nurturing the gardener as it is about producing abundance. Pull up a chair in your garden and create a space to sit and enjoy the fruits of your labor.

Regularly check the seedlings you direct seeded for pests and add preventatives or insect covering if necessary.

Deadhead flowers like zinnias, dahlias, and basil to promote more growth.

Don't forget about those unexpected late frosts; keep your row covers handy.

Interplant root veggies like carrots, radishes, or beets alongside your tomatoes. (Spinach and baby greens are great to interplant with your tomatoes as well.)

Prep your beds for summer plantings by removing weeds, loosening the soil, and top dressing with amendments.

Rotate crops from the early season to reduce pests and diseases and promote soil health and fertility.

Check irrigation systems for leaks and holes in the early season and fix any leaks or timers before planting out your warm-season crops.

Mulch plants to retain moisture; choose organic mulch options like wood chips, straw, or leaves.

Don't forget to keep track of your planting dates, your plants' progress and yields, as well as likes and dislikes. Now is the time to start evaluating whether you want to continue growing these crops again next season.

Continue to prune crops like tomatoes, cucumbers, and peppers weekly. Remove foliage for better airflow and to redirect energy towards fruit yield.

Continue to succession plant your peas, beans, tomatoes, and other warm-season varieties to ensure multiple harvests throughout the season.

Overwhelmed by the amount of food you are growing? Call your friends, family, or neighbors and plan a community gardening day. Weed together, harvest together, and plan a time to preserve the bounty. Sharing the responsibility and fellowship does wonders for the gardener.

Prune off the suckers on your tomatoes (P.S. You can root them in water or soil and create free tomato plants from your trimmings!)

Be diligent and keep harvesting; the more you harvest, the more your plants will produce. If you're overwhelmed by the bounty, freeze things like tomatoes for the winter.

Don't forget to thin the seedlings you direct sowed. Use radishes, beets, spinach, and carrots as microgreens to top salads, sandwiches, or omelets.

Clean your gardening basket, pruners, and harvesting supplies. I like to keep these in the same spot, making them easy to find when I need them.

Open the vents and windows on your greenhouses to allow airflow on warm days.

Install trellises and stakes for climbing plants like peas and cucumbers, or tomatoes and peppers.

Plan how you will preserve your garden bounty. Look for recipes, new and old, and start eating and using up last season's preserves. Make way for the new.

Direct seed those warm-weather-loving crops like corn, beans, and squash.

Stock up on preserving supplies. Whether you need canning jars and lids, or fermenting essentials like weights, salt, and Pickle Pipes®, now is the time to gather all those items before the big harvests roll in.

Monitor watering needs as the weather warms up; add in additional watering or water for longer periods of time. Water deeply at the root level and avoid foliar watering to prevent disease and sunburn on your plants.

Chore Tracking

I find that organizing my mid-season chores by the day of the week is helpful for staying on track. Maybe Wednesday is harvesting day and Thursday is preserving day on your farm. Use this chart to track your mid-season chores and remember what to do on each day of the week. It will be helpful as the season progresses to know what needs to be done when. Chores in the Any Day column can be done whenever you have a bit of extra time in your schedule.

MONDAY	TUESDAY	WEDNESDAY	THURSDAY

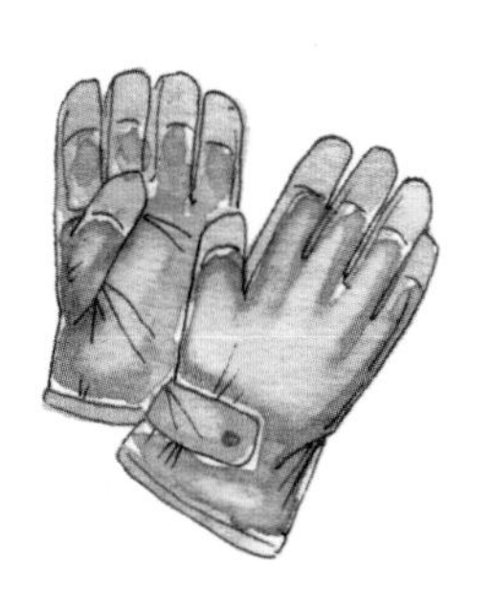

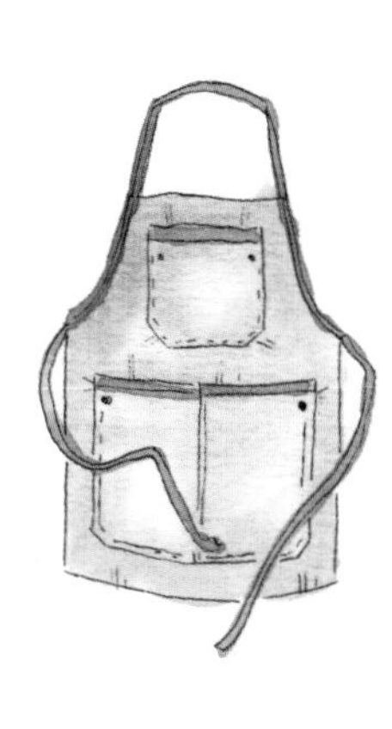

FRIDAY	SATURDAY	SUNDAY	ANY DAY

My Farm's Mid-Season Inventory: What We Have

SEEDS	SOIL	SOIL AMENDMENTS	TOOLS	EQ

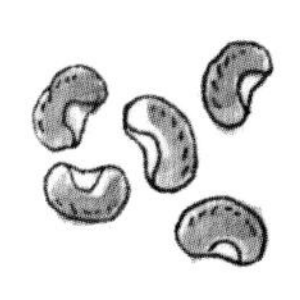

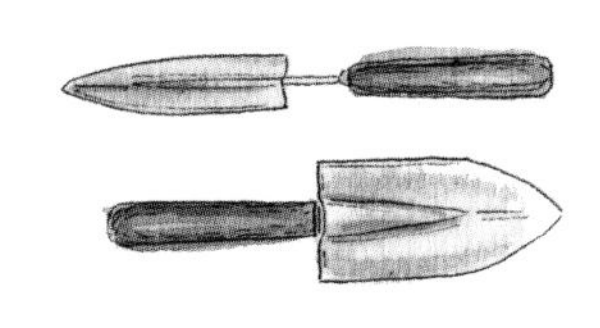

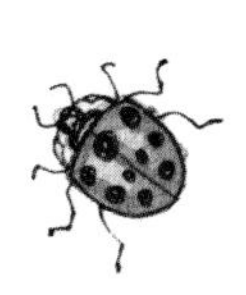

ENT	MULCH	HARVESTING SUPPLIES	PRESERVING SUPPLIES	ADDITIONAL ITEMS

Mid-Season Supply Needs: What We Need

SEEDS	SOIL	SOIL AMENDMENTS	POTS &

 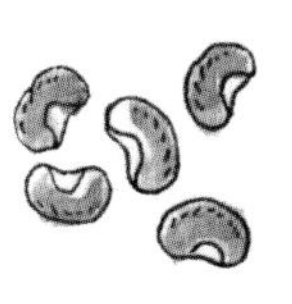  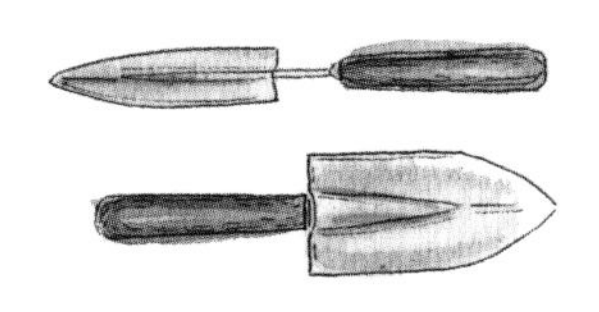  

CONTAINERS	TOOLS	IRRIGATION SUPPLIES	ADDITIONAL ITEMS

Mid-Season Expense Log

DATE	CATEGORY	VENDOR	ITEM PURCHASED

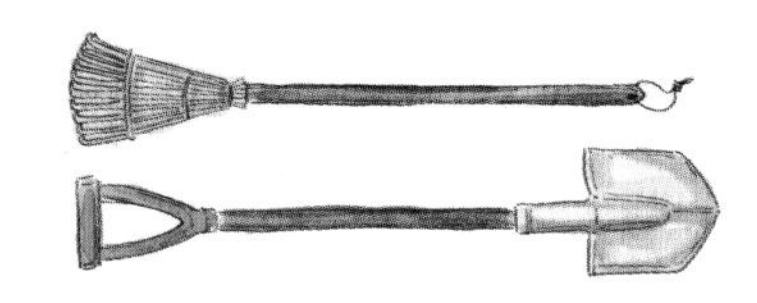

QUANTITY	TOTAL COST	PAYMENT METHOD	NOTES

Seed-Starting Schedule

Use this calendar to record your seeding dates, numbers of starts, how you planted and location in which you sowed them out. It's important to use pencil as these change as you lay out your gardens. You may need to erase and rewrite a few times until your final planting calendar is final.

TP = Transplant DS = Direct Seed

CROP	SEEDING DATE	SIZE TRAYS	# OF STARTS	TP OR DS	LOCATION	SUCCESSION PLANTING

CROP	SEEDING DATE	SIZE TRAYS	# OF STARTS	TP OR DS	LOCATION	SUCCESSION PLANTING

Herb-Harvest Tracker

Use this herb-harvest tracker as a tool to refine your harvest strategy and planning for upcoming seasons. Learn the most optimal herbs for your growing zone and their yields and plan their usage. Document your intentions to cultivate and harvest efficiently to ensure a bountiful harvest in the future.

HERB	HARVEST DATE	QUANTITY HARVESTED	CULINARY OR MEDICINAL USE

HERB	HARVEST DATE	QUANTITY HARVESTED	CULINARY OR MEDICINAL USE

Mid-Season Pest and Disease Tracker

DATE	PEST/DISEASE	PLANT AFFECTED	SYMPTOMS

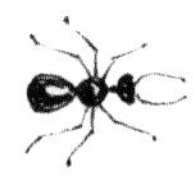

NTENSITY LEVEL	TREATMENT	PREVENTION	NOTES

Mid-Season Variety Reviews

CROP	VARIETY	LOCATION	PESTS	DISEASES	FLAVOR

PERFORMANCE	YIELDS	PROS/CONS	EXTRA NOTES

Mid-Season Variety Reviews

CROP	VARIETY	LOCATION	PESTS	DISEASES	FLAVO

PERFORMANCE	YIELDS	PROS/CONS	EXTRA NOTES

Planting Calendar Chart

CROP	INDOOR START DATE	TRANSPLANT DATE	DIRECT SEED DATE	SPACING

OIL REQUIREMENTS	SUN REQUIREMENTS	WATER REQUIREMENTS	NOTES

Planting Calendar Chart

CROP	INDOOR START DATE	TRANSPLANT DATE	DIRECT SEED DATE	SPACING

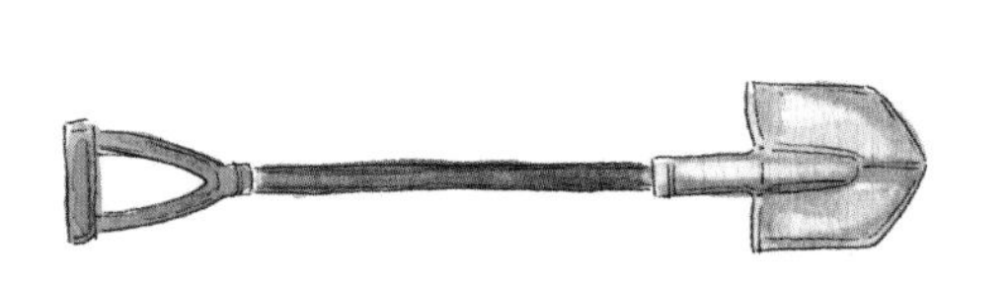

IL REQUIREMENTS	SUN REQUIREMENTS	WATER REQUIREMENTS	NOTES

Seasonal Transition: Mid-Season to Late Season

As we dive into the lively mid-season of gardening, pouring our love into nurturing our blossoming gardens, it's vital to get ready for the shift into the next phase. With the days stretching longer and warmer, it's almost time to turn our focus towards the late season. This marks a time of change in the garden, moving from caring for our thriving summer crops to getting ready for the winding down of summer and the prospects of a winter harvest.

"I grow plants for many reasons: to please my soul, to challenge the elements, or to challenge my patience, for novelty or for nostalgia, but mostly for the joy of seeing them grow."
—David Hobson

During mid-season, the garden is buzzing with activity. Seeds we planted are bursting into life, flowers are adding pops of color, and the first fruits of the season are making their appearance. It's a reward for all the careful planning and attention we've given. But as mid-season begins to wind down, we need to expect the summer crops to slowly fade, making space for a new growth cycle.

The late season beckons, signaling a shift toward getting set for the move from the summer plenty to the excitement of a winter harvest. As the summer crops gracefully take their leave, we start planning for the cooler months ahead. Now is the time to check the condition of our soil, getting it ready for the next planting spree. It's the perfect moment to order seeds for the winter garden, prep our greenhouse for some indoor planting, and maybe make some adjustments based on soil tests.

During this transition, let's pause and appreciate the bounty of our hard work during mid-season as we gear up for the late season. For me this shift is something I eagerly welcome. It's the final opportunity for those second summer crops and a chance to start preparing for a slower season ahead. The chores aren't as demanding, the cool mornings invite a gentler pace, and I know these last few months will be cherished throughout winter until the cycle begins anew, and we're once again surrounded by ripe tomatoes.

Month-by-Month Late-Season Checklist

Jill's Checklist

Month 1: Late-Season Management

- ☐ Maintain regular and consistent care for established plants.
- ☐ Monitor soil moisture and water deeply during dry spells. Mulch as needed.
- ☐ Continue harvesting and processing summer produce.
- ☐ Start preserving your abundance through canning, fermenting, freezing, or dehydrating.
- ☐ Implement strategies to reduce heat stress on crops (shade cloth, heat-tolerant varieties, mulch, etc.)
- ☐ Remove diseased plants and plant another succession.
- ☐ Evaluate financial goals and adjust.
- ☐ Research and plan for winter storage and preservation techniques.

Month 2: Harvesting and Preservation

- ☐ Preserve excess produce through canning, drying, or freezing.
- ☐ Monitor and manage late-season pests.
- ☐ Harvest and store seeds from open-pollinated plants.
- ☐ Begin transitioning annual beds to fall crops.
- ☐ Prune and shape perennials and fruiting plants.
- ☐ Collect soil tests to prepare for fall planting.
- ☐ Amend and fertilize the garden beds that you are cleaning out.
- ☐ Reflect on the summer successes and challenges for future improvements.

Month 3: Fall Plantings and Harvest Preparations

- ☐ Plant cool-season crops for fall and winter harvest.
- ☐ Begin cover cropping on vacant beds to enrich the soil.
- ☐ Divide and transplant perennials as needed.
- ☐ Harvest and store seeds from summer crops.
- ☐ Start scouting for fall pests and diseases (apply insect netting if needed).
- ☐ Clean out spent summer crops.
- ☐ Plant spring flowering bulbs.
- ☐ Start reviewing seeds and supplies you'll need for next year.

My Checklist

- ☐
- ☐
- ☐
- ☐
- ☐
- ☐
- ☐
- ☐
- ☐
- ☐
- ☐
- ☐
- ☐
- ☐
- ☐
- ☐
- ☐
- ☐
- ☐
- ☐
- ☐
- ☐
- ☐
- ☐
- ☐
- ☐
- ☐
- ☐

Late-Season Friendly Reminders

Regularly aerate the soil to improve its structure.

Check your irrigation systems for efficiency.

If you are limited in space, utilize vertical gardening to get the most out of your area.

Order seeds and supplies for your fall gardens.

Mulch with leaves, straw, or wood chips to retain moisture and deter weeds.

Fill out your planting calendar to anticipate when you will begin harvesting and preserving your bounty.

If you have empty beds, plant a cover crop of rye or clover to help prevent soil erosion.

Start ordering your seed and bulb catalogs.

Plan and organize your fall and winter garden.

If you're planning to overwinter bulbs or flowers, this is the right moment to assess your supply requirements, decide on the varieties you wish to grow, and place orders for bulbs, tubers, or corms.

Try growing a second round of summer crops from hybrid varieties. Some staples for me are cucumbers, radish, and quick baby greens.

Make a note of the varieties you plan to save seeds from. Mark them and make sure you're employing the correct techniques to preserve true seeds.

Continue succession planting for a continual harvest. Simply stagger your plantings every 7 to 21 days.

Consider companion planting. Interplant basil and marigolds for natural pest control.

Consider ordering more trellising clips and supplies if needed.

Continue planting out your successions of flowers for a continuous harvest. Sunflowers are one of my favorites to grow, needing no support, and they are great for open-field cultivation.

Keep detailed documentation of pests, diseases, likes, and dislikes for each variety you grow. This will help you decide what to grow the following year.

Order preserving supplies, such as canning jars or fermenting weights and salt. Clean your dehydrators or freeze dryers and prepare for the harvest.

Consider growing hybrids, especially if you want to plant later in this season and still ensure a yield.

Continue harvesting or deadheading flowers for continued growth throughout the season.

Continue direct seeding for beans, squash, and peas.

Continue monitoring your pH levels for optimal soil health.

Don't forget to keep up with your seasonal expenses and jot them down for tax time.

Rotate your crops from the previous season to help improve soil health.

Order or buy your seed-starting mix for the fall garden.

Increase your watering times or frequency, depending on your zone.

Make sure you have plenty of soil amendments on hand to apply as needed.

Clean and sanitize your seed-starting trays for the next season.

If you have not yet conducted a soil test, collect a sample and send it off for recommended amendment additions this season.

Plan your pest and disease management for the season. Order supplies if necessary.

If selling wholesale or at a farmers' market, make sure you have plenty of buckets and totes for harvesting and supplies for packing and selling.

Keep your garden shears and tools clean and sanitized to avoid spreading disease.

Order preserving supplies, such as canning jars or fermenting weights and salt. Clean your dehydrators or freeze dryers and prepare for the harvest.

Roll up the sides of your high tunnels to maintain good airflow and manage disease.

Toward the end of the season, start planting out root veggies, spinach, and greens for an early fall harvest.

Chore Tracking

I find that organizing my late-season chores by the day of the week is helpful for staying on track. Maybe Wednesday is canning day and Thursday is pull out old plants day on your farm. Use this chart to track your late-season chores and remember what to do on each day of the week. It will be helpful as the season progresses to know what needs to be done when. Chores in the Any Day column can be done whenever you have a bit of extra time in your schedule.

MONDAY	TUESDAY	WEDNESDAY	THURSDAY

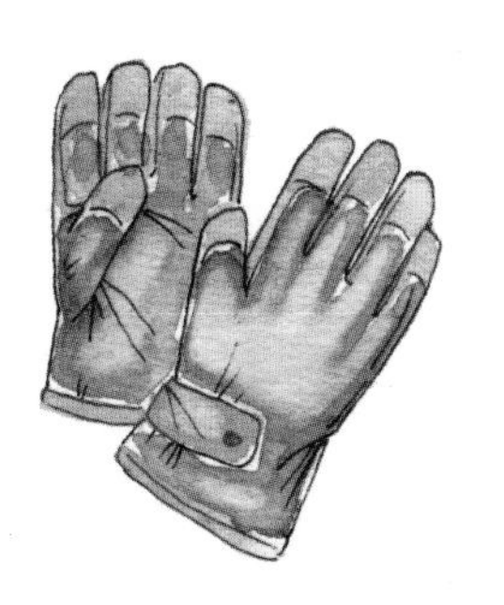

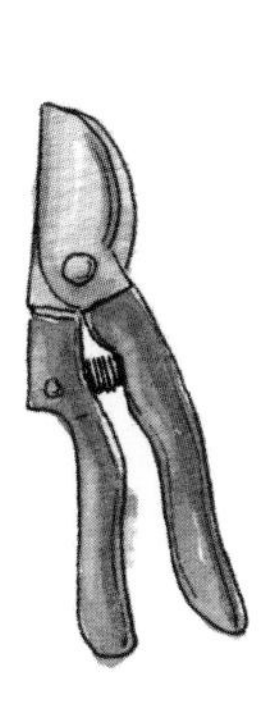
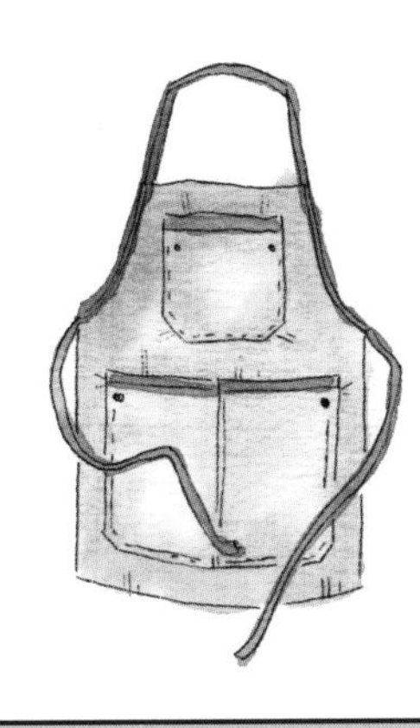

FRIDAY	SATURDAY	SUNDAY	ANY DAY

My Farm's Late-Season Inventory: What We Have

SEEDS	SOIL	SOIL AMENDMENTS	TOOLS	EQ

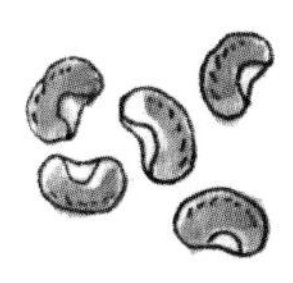

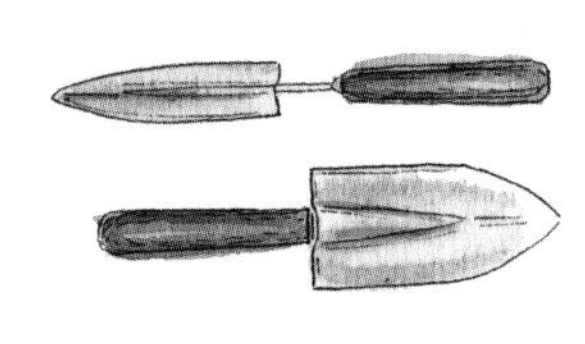

NT	MULCH	HARVESTING SUPPLIES	PRESERVING SUPPLIES	ADDITIONAL ITEMS

Late-Season Supply Needs: What We Need

SEEDS	SOIL	SOIL AMENDMENTS	POTS &

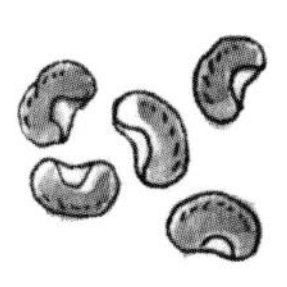

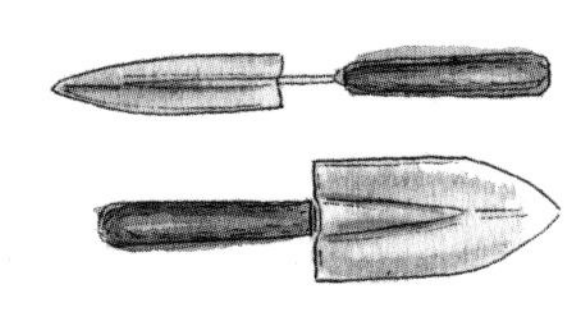

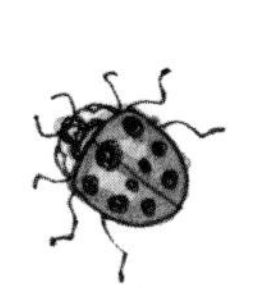

CONTAINERS	TOOLS	IRRIGATION SUPPLIES	ADDITIONAL ITEMS

Late-Season Expense Log

DATE	CATEGORY	VENDOR	ITEM PURCHASED

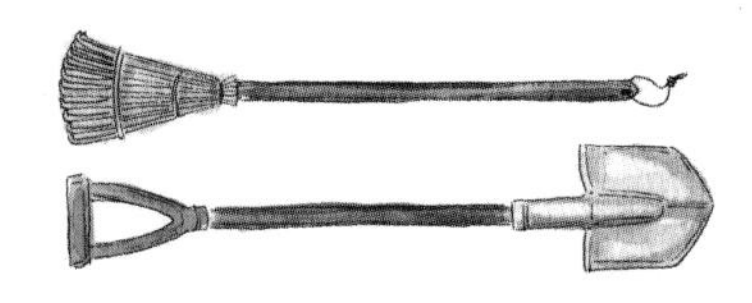

QUANTITY	TOTAL COST	PAYMENT METHOD	NOTES

Seed-Starting Schedule

Use this calendar to record your seeding dates, numbers of starts, how you planted and location in which you sowed them out. It's important to use pencil as these change as you lay out your gardens. You may need to erase and rewrite a few times until your final planting calendar is final.

TP = Transplant DS = Direct Seed

CROP	SEEDING DATE	SIZE TRAYS	# OF STARTS	TP OR DS	LOCATION	SUCCESSION PLANTING

CROP	SEEDING DATE	SIZE TRAYS	# OF STARTS	TP OR DS	LOCATION	SUCCESSION PLANTING

Seed-Saving Record

Secure the future of your garden with this Seed-Saving Chart. Record plant names, seed types, and quantities to ensure an ongoing growth cycle. Strategically store seeds for self-sufficiency, reducing the need to buy seeds. This valuable reference helps you track prolific plants, enabling efficient planning for future harvests.

CROP	VARIETY	SEED TYPE	QUANTITY HARVESTED	STORAGE LOCATION	HARVEST DATE

CROP	VARIETY	SEED TYPE	QUANTITY HARVESTED	STORAGE LOCATION	HARVEST DATE

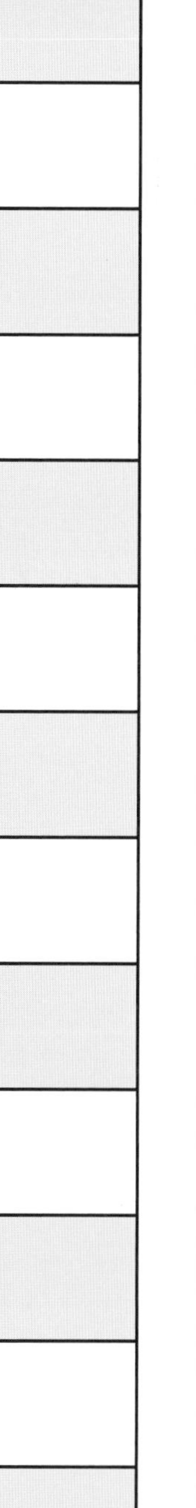

Late-Season Pest and Disease Tracker

DATE	PEST/DISEASE	PLANT AFFECTED	SYMPTOMS

  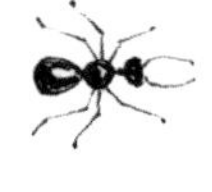 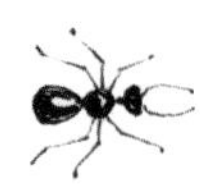   

INTENSITY LEVEL	TREATMENT	PREVENTION	NOTES

Late-Season Variety Reviews

CROP	VARIETY	LOCATION	PESTS	DISEASES	FLAVOR

PERFORMANCE	YIELDS	PROS/CONS	EXTRA NOTES

Late-Season Variety Reviews

CROP	VARIETY	LOCATION	PESTS	DISEASES	FLAVOR

PERFORMANCE	YIELDS	PROS/CONS	EXTRA NOTES

Planting Calendar Chart

CROP	INDOOR START DATE	TRANSPLANT DATE	DIRECT SEED DATE	SPACING

IL REQUIREMENTS	SUN REQUIREMENTS	WATER REQUIREMENTS	NOTES

Planting Calendar Chart

CROP	INDOOR START DATE	TRANSPLANT DATE	DIRECT SEED DATE	SPACING

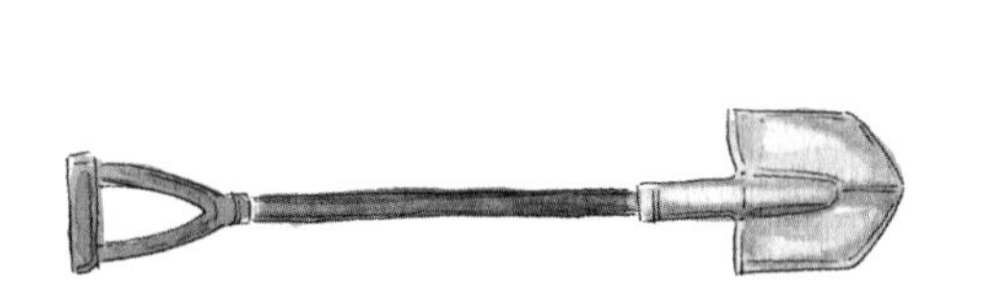

IL REQUIREMENTS	SUN REQUIREMENTS	WATER REQUIREMENTS	NOTES

Seasonal Transition: Late Season to Winter Season

"And all at once, summer collapsed into fall."
—Oscar Wilde

Congratulations, friend! We've successfully navigated the late season, and the much-anticipated shift in weather has finally arrived. Mornings now greet us with a refreshing crispness, signaling the departure of our heat-loving crops and the onset of a slower pace. For some, this marks the end of the gardening season until the next spring. But for many of us, it's a chance to embrace the cool, brisk winter and cultivate an array of frost-hardy plants.

To my surprise, winter has become my favorite time for gardening. It's difficult to put into words, but there's a unique allure to it. The garden transforms, and beauty manifests in mysterious ways. As we plant our brassicas and prepare for days of harvesting cabbage and making kimchi, there's a sense of satisfaction. If you are not growing winter crops, consider this a time for rest and rejuvenation, nourishing your mind, body, and soul.

Immerse yourself in every seed catalog within reach. Let your imagination wander to the upcoming spring season. Discover new varieties for your garden, contemplate a fresh layout, experiment with innovative techniques, and absorb knowledge during this restful period. Reflect on the successes and challenges of the past season, allowing them to guide your plans for the next gardening cycle.

Winter gardening has a different rhythm and pace, urging us to savor the experience. The sight of plants thriving in colder temperatures is a testament to nature's resilience and the gardener's dedication. As the days grow shorter and the nights longer, there's a sense of quiet satisfaction in nurturing a garden through the winter months. And as the snow blankets the landscape, the anticipation of spring and the promise of new growth keep our gardening spirits alive.

Whether you're tending to a winter garden or taking a well deserved break, cherish this transitional time. Embrace the change, learn from the past, and let the excitement for what's to come fuel your gardening journey.

Month-by-Month Winter Season Checklist

Jill's Checklist

Month 1: Winter Harvest and Cleanup

- ☐ Continue harvesting cool-season crops.
- ☐ Clean up and mulch garden beds for winter.
- ☐ Set up cold frames and row covers for an extended harvest.
- ☐ Store or process harvested produce for market or preservation.
- ☐ Prepare garden infrastructure for colder weather.
- ☐ Continue to cover-crop beds or pastures.
- ☐ Continue planting frost-hardy crops like lettuce, spinach, and chard.
- ☐ Plant garlic and shallots for next year's harvest.

Month 2: Winter Prep and Reflection

- ☐ Finish winterizing equipment and tools.
- ☐ Complete final garden cleanup, removing debris and dead plants.
- ☐ Apply organic mulch to protect the soil during winter.
- ☐ Plan and design any new infrastructure or expansions for the next year.
- ☐ Attend farm or garden conferences or webinars for inspiration for the upcoming year.
- ☐ Review farm finances and prepare year-end reports.
- ☐ Evaluate preservation goals.
- ☐ Reflect on the year's successes and challenges for continuous improvement.

Month 3: Rest and Dream for Next Year

- ☐ Rest and rejuvenate before the next growing season (you've earned it!).
- ☐ Review and assess the past year's successes and failures.
- ☐ Shop seed catalogs and plan next year's crops.
- ☐ Organize and maintain equipment, tools, and seeds.
- ☐ Develop a tentative planting schedule for the upcoming year.
- ☐ Set goals and objectives for the new year.
- ☐ Write a budget for the upcoming growing season.
- ☐ Dream big and plan wildly for the next growing season!

My Checklist

☐ ☐ ☐ ☐ ☐ ☐ ☐ ☐ ☐ ☐ ☐ ☐ ☐ ☐

☐ ☐ ☐ ☐ ☐ ☐ ☐ ☐ ☐ ☐ ☐ ☐ ☐ ☐

Winter Season Friendly Reminders

Start placing seed orders for the spring.

Take down trellising systems to prepare for the new year.

If growing in a greenhouse or high tunnel, be diligent in continuing to harvest and preserve throughout the season.

Order garden supplies like heat mats, soil, trays, and pots.

Clean and oil any garden tools before storing them. This saves time in the spring.

Don't forget to roll down the sides of your high tunnel to trap heat as the temperature begins to drop.

Place a heater in your well house to prevent your pump from freezing.

Take soil samples from each of your garden areas and send samples off. This will give you enough time to gather amendments and get your beds ready for spring plantings.

Manage the humidity in your greenhouses and high tunnels. Excessive moisture can lead to fungal diseases in your plants, so remove the row cover during the day and ventilate properly. If rolling the sides up on your tunnel during sunny winter days, be sure to roll them down mid-afternoon to ensure enough heat is trapped.

Plant winter-hardy cover crops to protect and enrich your soil during the winter months. They act as a natural blanket for your garden.

Water matters this time of year. Be sure to water deeply before a hard freeze to keep plants moist before a dry winter spell.

Before a hard freeze, drain your water lines to prevent pipes from bursting.

Work on farm improvements. Now is a great time to tackle a new project or reconfigure your current structures or layouts.

Mulch around your winter plants (perennials) to help retain moisture, regulate the soil temperature, and prevent any unexpected weeds that might pop up.

Use the winter down time to plan next year's garden layout. Dream big, map out, create vision boards, and anticipate what the new year will look like.

Remember those winter supplies we had you stock up on in the late season? Now is the time to pull those out and keep your winter plants nice and cozy.

Test your soil pH during the winter so that you can adjust your soil before spring planting.

Adjust your irrigation frequency in the winter. Water less frequently but deeply at the base.

If growing flowers, order floral supplies for the spring, like trellising, rubber bands, paper wraps, pruners, etc.

Continue harvesting winter veggies like kale, carrots, brassicas, and more.

Inspect and repair any damaged plastic or insulation in your greenhouse or high tunnel. A well-insulated space is critical for overwintering plants.

If starting seeds indoors, begin to prepare your seed-starting areas.

On sunny days, remove the row cover to allow better airflow for your plants.

Don't forget to record your winter observations, yields, and variety reviews.

Start seeds indoors for those early spring crops. Starting seeds ahead of time allows you a head start on the growing season.

Drain and insulate pipes and cover your outdoor faucets to prevent damage in freezing temperatures.

Research any new varieties or trellising techniques you might want to use in the next season.

Insulate your root vegetable beds (carrots, radishes, beets, rutabagas, turnips) with a thick layer of straw or mulch to prevent freezing, then harvest from this bed all winter.

Continue planting successions of winter-hardy varieties for early spring harvest.

Avoid overhead watering in freezing temperatures; wet foliage can lead to frost damage on your plants. Water in the morning when temperatures are a bit higher.

Turn your compost pile regularly during the winter. This will help keep the microbial activity going, even when it gets colder.

Chore Tracking

I find that organizing my winter chores by the day of the week is helpful for staying on track. Maybe Wednesday is canning day and Friday is seed inventory day on your farm. Use this chart to track your winter chores and remember what to do on each day of the week. It will be helpful as the season progresses to know what needs to be done when. Chores in the Any Day column can be done whenever you have a bit of extra time in your schedule.

MONDAY	TUESDAY	WEDNESDAY	THURSDAY

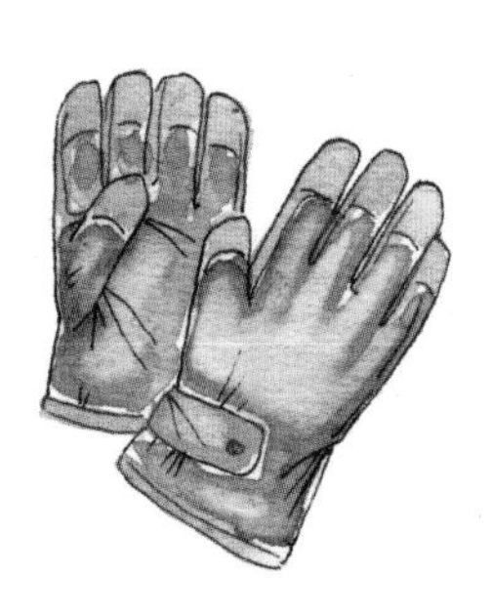

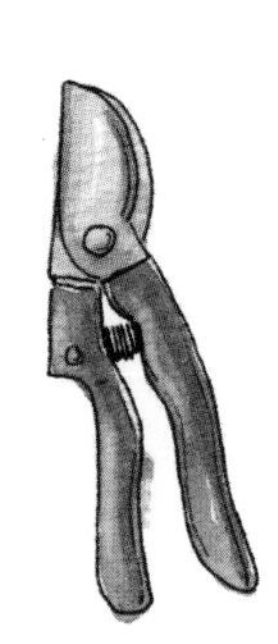
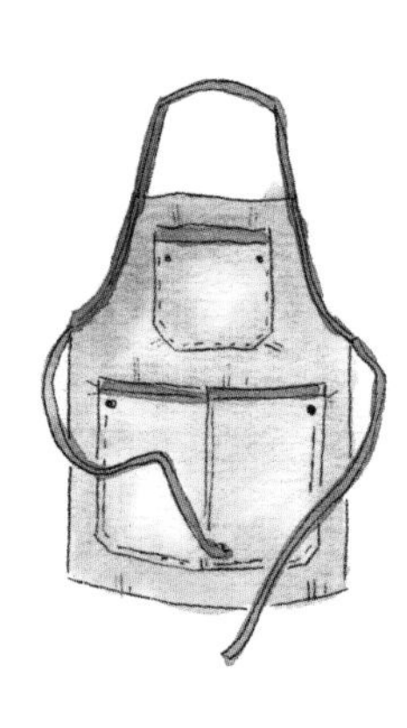

FRIDAY	SATURDAY	SUNDAY	ANY DAY

High Tunnel Management

High tunnels are invaluable for extending the growing season and protecting your crops. An effective management and maintenance plan will ensure you make the most of them and achieve successful harvests throughout the year. Let's break down a comprehensive plan that covers aspects of growing in a high tunnel.

Shade Cloth Management

Shade cloth is a valuable tool for regulating temperature and light within the high tunnel. The percentage on the shade cloth indicates the amount of light it blocks. Lower percentages (e.g., 30 percent) allow more light transmission, ideal for cooler days. Higher percentages (e.g., 70 percent) block more light and are suitable for hot, sunny days.

Putting on Shade Cloth

Late Spring to Early Summer: As the weather warms up, typically in late spring or early summer, it's a good time to put on shade cloth. This helps in preventing excessive heat and sun exposure, protecting plants from scorching and stress.

Taking off Shade Cloth

Fall: As the weather starts to cool down in early fall, and the intensity of the sun decreases, it's advisable to remove the shade cloth. This allows plants to receive more sunlight, which becomes especially important for crops that thrive in cooler temperatures.

Frost Dates

Setting frost dates is crucial for planning and protecting crops. Adjusting the dates two to four weeks earlier and later than the typical outdoor growing season accounts for the warming effect of the high tunnel and helps plan plantings and harvests accordingly.

Fertilization Schedule

Fertilization is vital to plant growth. In spring, plants enter the active growth phase and require a balanced fertilizer rich in nitrogen, phosphorus, and potassium. In fall, a fertilizer higher in phosphorus and potassium promotes root development and overall plant hardiness.

Irrigation Plan

Proper irrigation is essential for plant health. During the peak hot seasons, aim for approximately 45 minutes of watering per day. In mid-winter, reduce the frequency to once a month, adjusting based on plant needs and weather conditions.

Growing in-Ground vs. Raised Beds

Choosing between growing directly in the ground within the tunnel versus utilizing raised beds involves considering factors such as drainage, soil health, and ease of management. While in-ground growth allows plants to access the native soil, raised beds offer better control over soil quality, drainage, and warmth.

Crop Rotation

Crop rotation is a preventive measure to avoid soil-borne diseases and improve soil fertility. Moving crops within the high tunnel annually helps break pest cycles and ensures balanced nutrient uptake.

Pest and Disease Management

You should use Integrated Pest Management (IPM) strategies, such as biological controls, cultural practices, and organic treatments. Regularly monitor for pests and diseases, and promptly address any issues to prevent widespread infestations.

Ventilation and Temperature Control

Proper ventilation regulates temperature and humidity inside the high tunnel. Adjust vents or fans to maintain the optimal temperature range for the specific crops you are growing.

Optimizing Sunlight

Placing taller crops on the north side and shorter ones on the south side optimizes sunlight exposure. Adjust planting beds accordingly to ensure each plant receives adequate light.

Trellising and Support Systems

Tall or vining crops benefit from trellising systems, allowing for better airflow and light penetration. Support systems reduce pest damage and make harvesting easier. You can use the top of your high tunnel as support for certain trellising techniques.

Crop Selection and Succession Planting

Choose crops suitable for your high tunnel and plan for successive plantings for a continuous harvest. Start with cool-season crops, then transition to warm-season ones as temperatures rise inside the tunnel.

Record Keeping

Detailed records help you analyze past performance and plan future activities. Track planting dates, varieties, yields, fertilization schedules, and any pest or disease occurrences. Use this information to make informed decisions.

Equipment Maintenance

Regularly maintain and service gardening equipment to keep it working well. Check irrigation systems, fans, heaters, and any other equipment routinely to prevent disruptions during crucial times.

By following these tips for high tunnel management and maintenance, you'll be equipped to optimize your growing space and reap a successful and bountiful harvest all year round. Adapt and customize this plan to suit your specific growing goals and circumstances.

My Farm's Winter Inventory: What We Have

SEEDS	SOIL	SOIL AMENDMENTS	TOOLS	EQU

 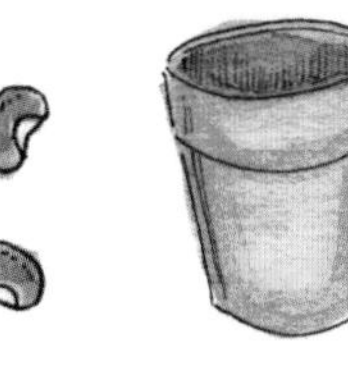 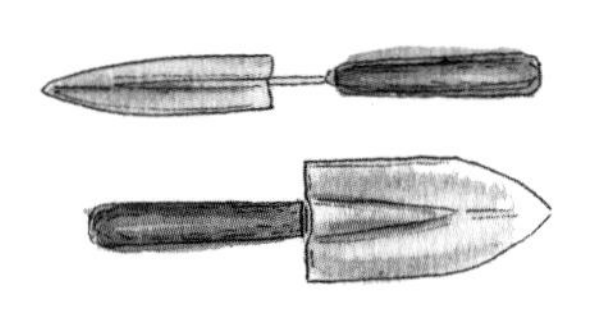  

ENT	MULCH	HARVESTING SUPPLIES	PRESERVING SUPPLIES	ADDITIONAL ITEMS

Winter Season Supply Needs: What We Need

EQUIPMENT	TOOLS	FERTILIZERS & AMENDMENT

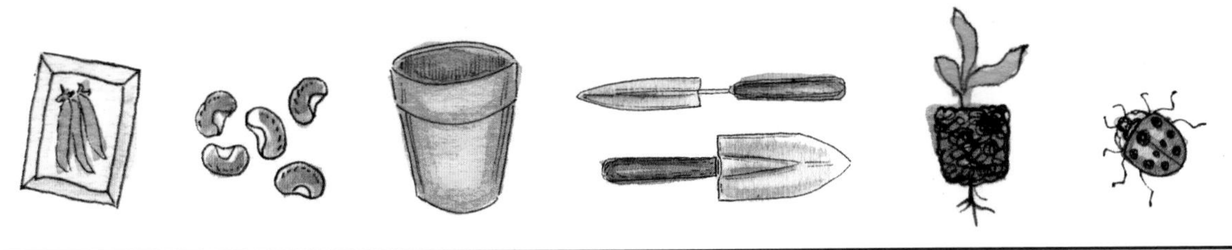

TOOLS	SOIL & MULCHES	ADDITIONAL ITEMS

Winter Expense Log

DATE	CATEGORY	VENDOR	ITEM PURCHASED

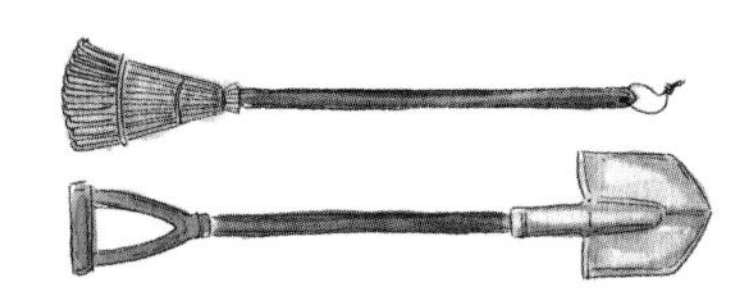

QUANTITY	TOTAL COST	PAYMENT METHOD	NOTES

Favorite Winter Varieties

CARROTS, NAPOLI

One of my top reasons for winter growing is the remarkable transformation in the flavor of vegetables. After that initial frost, everything takes on a sweeter note when you grow them yourself, unlike those boring vegetables you can buy at the store. Carrots stand out for me. While we cultivate them in both spring and winter, after that first frost, they turn sweet, almost like candy. Napoli is fantastic for storage and consistently delivers uniform carrots, which is exactly why I've been dedicated to growing it for years.

Tip: Napoli carrots are perfect for winter growing. Sow in well-prepared soil and keep consistently moist. Harvest when carrots reach the desired size.

RADISHES, BRAVO

I could spend ages raving about everything I love about this variety, but let's keep it short and sweet. Firstly, it's a stunner, boasting the most vivid purple hue all the way through. These radishes are massive, and unusually, the larger size works to their advantage. Unlike most radishes that can get woody when big, this daikon variety is meant to be substantial. They hold up beautifully in the fridge; I've kept mine for up to a year. This makes them perfect for a single large planting, yielding harvests throughout the season. The flavor is bold yet not too spicy, making it an ideal candidate for ferments or as a raw snack or garnish. In my book, everyone should give these a shot at least once.

Tip: Directly sow seeds in well-drained, loose soil. Plant in early fall for a fall and winter harvest. Provide consistent moisture, especially during dry spells, to prevent woody texture. Harvest when radishes reach maturity, usually around 45 to 60 days after planting.

KOHLRABI, KOSSAK

Kossak has a tendency for reaching impressive sizes—really huge. In some years, when I unintentionally missed a harvest, I stumbled upon some larger than my head. Although not the ideal size for picking, it did give everyone a good laugh. When harvested young, however, these kohlrabies are perfect for storage. They can endure for months in the fridge, giving you a winter-long supply. Our favorite trick is fermenting them, adding a nice burst of flavor to our winter soups.

Tip: Directly sow Kossak kohlrabi seeds in well-drained soil. Harvest when bulbs are young and tender for the best flavor and texture.

BEETS, BORO

I've been growing this variety primarily for its impressive resilience in cold temperatures. Apart from being cold-resistant, it offers a fantastic, earthy beet flavor. Even when the foliage starts to wither, I continue to store the beets in the soil, harvesting them as needed.

Tip: Boro beets are hardy and can withstand cold temperatures. Plant in loose, well-drained soil and harvest young for the best flavor and tenderness.

BROCCOLI, IMPERIAL

If you haven't caught on, brassicas truly steal the spotlight in the winter garden, and Imperial is no exception. This broccoli variety produces large, beautiful heads, as well as abundant side shoots, ensuring you get the maximum yield. Not to mention, it's a reliable producer even in colder temperatures, and when paired with row cover, it keeps delivering throughout the entire winter season.

Tip: Plant Imperial broccoli in nutrient-rich soil with good drainage. Harvest the central head first, and side shoots will continue to develop for an extended harvest.

BRUSSELS SPROUT, DAGAN

In my opinion, there's nothing quite like Brussels sprouts in the heart of winter. Maybe it's because we exclusively grow them during this season, adding a special appreciation to each bite. But you can notice the difference when they're grown in cooler weather. The Brussels sprouts take on a slightly sweet undertone, noticeably milder, and they're delicious whether enjoyed raw, charred, steamed, or tossed into a stir fry.

Tip: Dagan Brussel sprouts are cold tolerant. Start seeds indoors and transplant in late summer. Harvest from the bottom up as the sprouts mature.

SPINACH, AUROCH

While many associate greens with spring and summer, my love for growing them, especially spinach, extends into the winter. Spinach is renowned for its excellence as a winter green: simple, low-maintenance, and capable of thriving when directly seeded. It continues to produce even as temperatures drop. Enjoy it in soups, stews, smoothies, and protein bowls throughout the winter season.

Tip: Auroch spinach is a versatile winter green. Sow seeds in early fall for a late fall or early winter harvest. Harvest the outer leaves for continuous production.

CHINESE CABBAGE, BILKO

Chinese cabbages are typically sown in late summer for a fall harvest, but we've taken a different approach. In our high tunnel, we plant multiple successions and use insect netting in the fall, and row cover during the winter. This unconventional strategy has proven to be highly successful for us. Specifically, Bilko, one of our favorite varieties, has consistently delivered stunning, uniform heads with a delightfully mild and sweet flavor, a perfect companion for crafting delicious kimchi.

Tip: Sow Chinese cabbage seeds in late summer for a fall harvest. Provide consistent moisture and consider using row covers to protect against pests.

KALE, REDBOR

You may have picked up on a theme—I'm all about those deep, vibrant colors in the winter garden, and this kale fits the bill. Its beauty is a constant source of amazement. Flourishing in cold temperatures, it's an ideal candidate for winter growing. Plus, with every frost, its flavor just keeps getting sweeter and sweeter.

Tip: Redbor kale adds color to winter gardens. Plant in well-drained soil and harvest young leaves for salads or mature leaves for cooking.

KOHLRABI, PURPLE VIENNA

While Purple Vienna may not have the same storage longevity as Kossak, it has nonetheless claimed a spot in my winter garden for several years. Its allure lies in both its striking beauty and delightful flavor. This mild variety thrives throughout the winter season, adding not only taste but also a visually pleasing element to the garden.

Tip: Purple Vienna kohlrabi is known for its vibrant color. Plant in full sun and provide consistent moisture. Harvest when bulbs are 2 to 3 inches (5 to 8 cm) in diameter.

CAULIFLOWER, PURPLE MOON

I've spent several years cultivating Purple Moon, and each time, it continues to amaze me. Not only does it have a striking appearance, retaining its beautiful color even through the cooking process, but it also brings a milder and sweeter flavor to the table. What's even better, it shows great resilience to mold and disease, especially during the winter months.

Tip: Purple Moon cauliflower thrives in cool weather. Plant in fertile soil and blanch the heads by tying the leaves over the developing curds to maintain their purple color.

CAULIFLOWER, EARLISNOW

Growing in a high tunnel or under row covers allows you to push the boundaries of what you can cultivate. I've been experimenting with Earlisnow under protective covers for 2 years, and even before that, utilizing row covers, and it consistently shines. This cauliflower variety produces substantial heads, making it perfect for creating cauliflower rice, fermenting, or freezing florets. It has effortlessly become one of our go-to winter staples.

Tip: For early harvest, start Earlisnow cauliflower indoors and transplant seedlings after the last frost. Provide well-drained soil and consistent moisture.

CABBAGE, TIARA

Tiara is one of my go-tos in the garden. I've grown it in both spring and winter. While it is intended to be a more personal-sized cabbage, we have let ours become quite large before harvesting, and it has a remarkable resistance to splitting.

Tip: Plant Tiara cabbage in well-drained soil with ample sunlight. Consistent watering is essential for optimal head formation. Harvest when heads are firm and dense.

RUTABAGA, HELENOR

Although rutabaga is often associated with fall, I've been growing this variety throughout the winter for several years under row covers. Rutabagas, often an overlooked gem, serve as an excellent substitute for starch. We use them in place of potatoes, and they provide a rich, smooth, and creamy alternative. Storing them in the soil for harvesting as needed makes them a resilient winter vegetable. They are a fantastic addition to any winter garden.

Tip: Start Helenor rutabaga seeds in late summer for a fall harvest. Plant in fertile soil and provide consistent moisture. Harvest when roots are 4 to 5 inches (10 to 13 cm) in diameter.

KALE, WINTERBOR

Winterbor has consistently been my top kale pick. It continues to produce abundantly throughout the winter, ensuring a continuous harvest. Thanks to our relatively mild winters, we typically skip covering our kale. The frost, in turn, adds a delightful, sweet flavor to it.

Tip: Winterbor kale is cold-tolerant and can withstand winter conditions. Plant in full sun and harvest leaves as needed, allowing the plant to continue producing.

Seed-Starting Schedule

Use this calendar to record your seeding dates, numbers of starts, how you planted and location in which you sowed them out. It's important to use pencil as these change as you lay out your gardens. You may need to erase and rewrite a few times until your final planting calendar is final.

TP = Transplant DS = Direct Seed

CROP	SEEDING DATE	SIZE TRAYS	# OF STARTS	TP OR DS	LOCATION	SUCCESSION PLANTING

CROP	SEEDING DATE	SIZE TRAYS	# OF STARTS	TP OR DS	LOCATION	SUCCESSION PLANTING

Winter Storm and Weather Event Tracker

Use this chart to document storm dates, types, and snow- or rainfall amounts, and assess damage to guide your winter farming strategy. Whether your plants face blizzards or freezing rain, this essential tool ensures resilience against the unpredictable forces of nature. This record is a great way to stay prepared and proactive for winter based on winter's past.

DATE	STORM TYPE	AMOUNT OF RAIN OR SNOWFALL	DAMAGE ASSESSMENT

DATE	STORM TYPE	AMOUNT OF RAIN OR SNOWFALL	DAMAGE ASSESSMENT

Winter Pest and Disease Tracker

DATE	PEST/DISEASE	PLANT AFFECTED	SYMPTOMS

     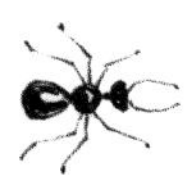 

INTENSITY LEVEL	TREATMENT	PREVENTION	NOTES

Winter Variety Reviews

CROP	VARIETY	LOCATION	PESTS	DISEASES	FLAVOR

PERFORMANCE	YIELDS	PROS/CONS	EXTRA NOTES

Planting Calendar Chart

CROP	INDOOR START DATE	TRANSPLANT DATE	DIRECT SEED DATE	SPACING

OIL REQUIREMENTS	SUN REQUIREMENTS	WATER REQUIREMENTS	NOTES

Planting Calendar Chart

CROP	INDOOR START DATE	TRANSPLANT DATE	DIRECT SEED DATE	SPACING

OIL REQUIREMENTS	SUN REQUIREMENTS	WATER REQUIREMENTS	NOTES

Seasonal Transition: Winter Season to Postseason

"Everything that slows us down and forces patience, everything that sets us back into the slow circles of nature, is a help. Gardening is an instrument of grace."
—May Sarton

Now that our gardens are taking a cozy winter nap (or maybe you're still celebrating your winter harvest victories), it's time to switch gears and dive into the exciting world of planning for the next growing season!

Throughout our gardening adventures, I've been waving the flag for planning and being ready for whatever Mother Nature throws our way. Well, guess what? Now's the time to dig out those notes you scribbled during the sunny days of spring and summer, kick back, and start dreaming about the glorious garden you're going to have in the spring.

Honestly this is one of my absolute favorite times to plan. There's way less stress in the air, and if you've been the diligent note-taker that I know you are, this transitional season will feel like a breeze. Take a moment to reflect on your previous growing year: which crops were high producers and which ones struggled a bit with pests, disease, or production?

Oh, and don't forget to peek at last season's prices, and make sure you adjust your budget accordingly. Don't forget to make sure your go-to supplier still has your favorite varieties in stock too.

Now's also the perfect time to spice things up a bit and explore new suppliers and new varieties for your farm or garden but be mindful of adding in too many new varieties as they may have different trellising support or growing needs.

Let your imagination run wild as you set some goals and expectations for the upcoming year. What do you want to achieve? What are you hoping to get out of your gardens this upcoming year? It's your garden. Dream big! Grab a cup of something warm, snuggle up with your gardening planner, and let the planning party begin. The seeds of an abundant gardening season are sown in these moments of reflection and anticipation.

Soil pH Adjustments

Before we can start adjusting the pH in our soil, we must first understand what pH is and why it's important for our plant's growth.

Soil pH is a measure of the acidity or basicity of soil. Soil pH is a key characteristic to make informative analyses regarding soil characteristics both qualitatively and quantitatively. Wikipedia defines pH as "the negative logarithm of the activity of hydronium ions in a solution."

Knowing the pH level is essential because some plants prefer more acidic soil while others prefer more alkaline. When you know this, you can make the appropriate pH adjustments to ensure each plant is getting the nutrients it needs from the soil. If you aren't sure of your soil's pH levels, then your plants could be getting too much or too little of something, and it will affect the plants' growth.

Acidic soil has a lower pH, while alkaline soil has a higher pH. This is because acidophilic microbes dominate in acidic soils. These microbes are great at breaking down the organic matter in soil, therefore, allowing nutrients like nitrogen and phosphorus to be released. Alkalophilic microbes predominate in those alkaline soils. They also will decompose organic matter but as it decomposes, they will start to release nutrients like calcium, and magnesium.

Let's cover some pH adjustments you can make to your soil and the effects they have. Some of the most common additives include the following:

Limestone: There are a few different types of lime you can add to your soil, including calcitic lime, which can be bought as ground lime, pulverized, or pelletized. Then there is also dolomitic lime. Calcitic lime is sourced from deposits primarily of calcium carbonate, while dolomitic lime is derived from deposits containing a mixture of calcium carbonate and magnesium carbonate, elevating the magnesium levels. You would add these to raise the pH levels in your soil, typically in the fall. When you apply lime to your soil, it will react with your soil's acidity, therefore neutralizing any excess hydrogen ions and raising your pH levels.

Sulfur: The most common sulfur to add to your soil is elemental sulfur. You can also add aluminum sulfate, which is an oxidized form of sulfur. You would add sulfur to lower your soil pH level in alkaline-dominant soils. Typically this is applied in the fall after you've conducted a soil test. When sulfur is applied, it reacts with the soil and then creates sulfuric acid, which will then release hydrogen ions and decrease the pH of the soil.

Wood Ash: Wood ash is a common solution to raise the pH levels in your soil; it also adds potassium and calcium. The key to adding wood ash is to add it in moderation so it doesn't have an adverse effect on your soil. Typically this is applied in the spring or fall season. Wood ash contains alkaline components that help balance out your soil's acidity.

While there are certainly more options than these, these are the most commonly used and readily available for many gardeners.

pH Adjustments

DATE OF SOIL TEST	pH MEASUREMENT	LOCATION	AMENDMENT ADDED	AMOUNT ADDED	FOLLOW UP pH 6 MONTHS TO A YEAR LATER

Types of Cover Crops to Plant

Cover crops are vital for soil health and fertility. Before I recommend cover crops, let's chat about what cover crops are and why they are so important.

A cover crop is not planted to harvest and eat, but to enrich and protect bare soil. Cover crops reduce soil erosion and improve soil fertility and quality by breaking down their nutrients.

Cover crops can be an important asset to your farm in many ways. Perhaps you grow only seasonally, leaving beds bare in winter. Instead planting a cover crop will feed the soil and help ensure life is thriving even in the off months. Here are just a few of the many benefits of planting and incorporating cover crops on your farm:

- ☐ They improve the soil by absorbing and storing nutrients, preventing them from washing away. As the cover crops decompose over time, they will release all of those nutrients back into the soil, building and improving fertility.
- ☐ Cover crops are also a great way to fight weeds. If native grasses are invading, your cover crop can outcompete the native weeds thereby reducing the weeds.
- ☐ Cover crops reduce soil compaction and help with soil erosion caused by wind and water. Some cover crops with deep root systems are great for breaking up hard soil and allowing air and movement.

These are just a few of the ways that planting cover crops can improve your soil and build soil health and fertility for a bountiful harvest. Now that we understand the value of cover crops for our soil health, let's explore some of my favorites.

Spring Cover Crops

Oats

Oats are a great spring cover crop because they grow so fast and can be planted with other vegetables. Oats can add a quick source of organic matter, which makes it a great choice if you're late planting a cover crop but still want those added benefits.

Red Clover

Red clover has a quick growing root system that helps break up compacted soil and prevent soil erosion. It's also a nitrogen source, meaning it will produce its own nitrogen and feed the soil when cut.

Summer Cover Crops

Field Peas

Field peas offer a range of benefits when planted as a cover crop, such as fixing nitrogen. Field peas, like other legumes, can fix nitrogen from the atmosphere with the help of symbiotic bacteria in their root nodules. Their deep root systems help break up compacted soil and add to the soil structure. They also help suppress weeds, prevent soil erosion, and reduce pests and disease.

Kale

Kale is great for aerating the soil due to its long taproot, which also helps prevent erosion. Its big, leafy greens help suppress weed pressure.

Fall Cover Crops

Austrian Winter Peas

Peas will help fix nitrogen in the soil, combat erosion and help suppress weeds. When growing peas, make sure you plant them early enough to become established before the frost.

Hairy Vetch

Hairy vetch is great for the fall and winter for its winter hardiness, and ability to fix nitrogen and suppress weeds.

Winter Cover Crops

Winter Rye

The dense rye is great for winter growing as it provides coverage for the soil against the harsh winter conditions.

Wheat

Winter wheat is frost hardy and will help protect the soil during the winter conditions, along with providing N-P-K.

Cover Crop Planting Chart

COVER CROP	PLANTING DATE	LOCATION	GROWTH

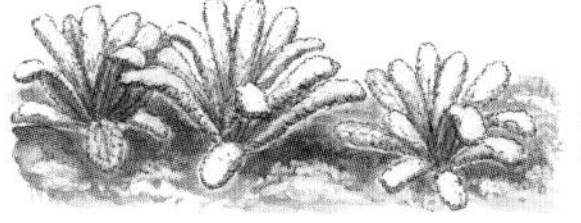

PERIOD	PURPOSE	DATE TURNED OR CUT	NOTES

Winterizing Your Farm

Winterizing your farm may not be the most thrilling task in your yearly routine, but it's undeniably one of the most crucial. As the winter season sets in, dedicating time to these essential chores ensures a smooth transition and sets the stage for a successful growing year when spring arrives. To set yourself up for success, I highly encourage you to use the Winterizing Chore Chart on the next pages to stay organized and on track with your tasks.

If you're uncertain about the necessary steps for winterizing your farm, don't worry. In this section, I'll provide insights into the winter chores we undertake on our farm. My husband, Nathan, and I share distinct winterizing responsibilities to keep our farm cozy in the colder months.

Our first chore involves cleaning and clearing the beds from the fall season. While our high tunnel maintains produce throughout winter, our cottage and kitchen gardens, along with raised beds, require attention. We remove any dead plants, compost them, and either plant cover crops or amend and mulch the beds. We cover in-ground beds with silage tarps secured with sandbags to inhibit weed growth in early spring, preventing water runoff and soil erosion in areas designated for spring planting.

Managing irrigation is a pivotal task. For gardens without irrigation systems, disconnecting and storing the water hose prevents freezing. For those with irrigation in place, it's crucial to unhook and drain the water lines. Leaving water in the lines during winter can lead to freezing and cracking, resulting in additional work and expense come spring. If your farm relies on a well, use a heater to prevent freezing.

Equally important is cleaning, sanitizing, and storing equipment and tools to prevent rusting. Regular equipment maintenance is a crucial aspect often overlooked. Taking the time to clean, lubricate, and properly store your tools during the winter not only prevents rusting but also ensures that they are in optimal condition when you need them for the busy spring season.

Mulching is another key chore during winter. Don't forget to mulch perennial plants and secure row covers for those still growing. In regions with heavy snowfall, make sure you are clearing snow off row covers to prevent collapsing and keep your plants safe.

For farms with infrastructure, winterizing takes center stage. Begin by cleaning the interior and stowing pots, tools, and equipment. Inspect and repair plastic, addressing holes promptly. Using high-quality plastic tape for repairs prevents the need for total replacement. Attend to any other repairs in the plastic or irrigation system to ensure proper functioning throughout winter.

Speaking of infrastructure, preparing your greenhouse or high tunnel for winter is crucial. Clean the interior thoroughly, removing any debris or leftover plant material. Inspect the structure for any damages and promptly repair them. Pay special attention to the plastic covering; patch any holes or tears using appropriate materials. Reinforcing weak points with additional support can prevent issues during heavy snowfall.

Winter also provides an ideal opportunity to clean and sanitize pots and trays. It's important to secure a supplier or stock up on soil for the upcoming spring. We prefer organizing our seed-starting greenhouse during winter, for a seamless transition when seed-starting season arrives. Check heat mats, grow lights, and any other equipment needed for seed starting. Replace or reorder any nonfunctioning items to guarantee a well-equipped and efficient setup.

I hope this shines light on how winterizing your farm involves a multifaceted approach encompassing bed preparation, irrigation management, equipment maintenance, and infrastructure care. By diligently addressing these tasks, you not only protect your farm during the winter months but also lay the groundwork for an abundant growing season ahead.

Winterizing Chore Chart

TASK	SUPPLIES NEEDED	RESPONSIBLE PARTY

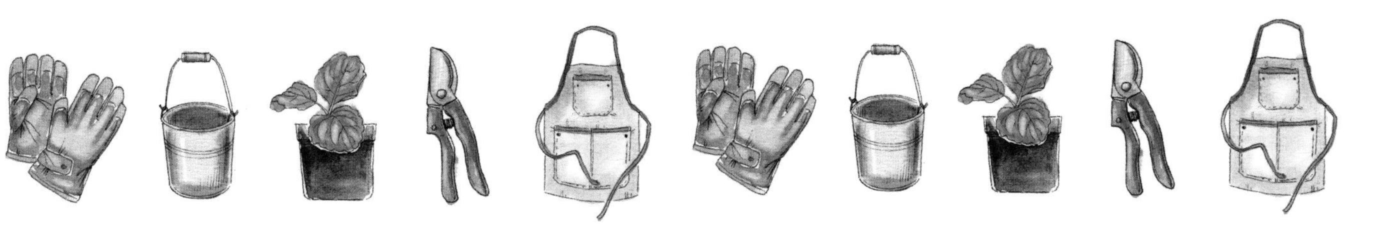

FREQUENCY	URGENCY	NOTES

End-of-Year Crop Evaluation

CROP	VARIETY	TOTAL YIELD	QUANTITY OF PLANTS	PESTS & DISEASES

CHALLENGES	SUCCESSES	PRESERVING	NOTES

End-of-Year Crop Evaluation

CROP	VARIETY	TOTAL YIELD	QUANTITY OF PLANTS	PESTS & DISEASES

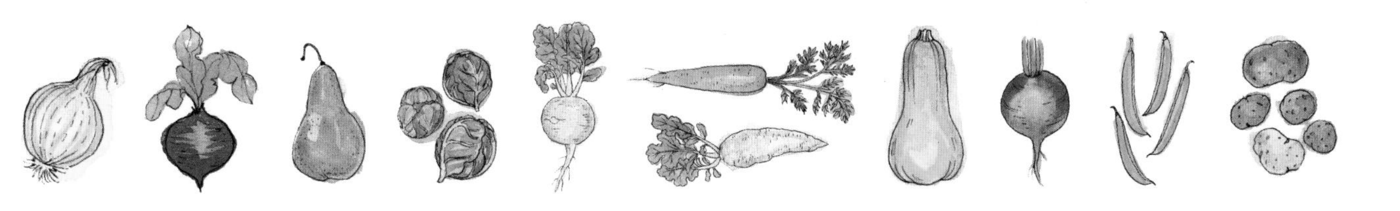

CHALLENGES	SUCCESSES	PRESERVING	NOTES

Canned Foods Inventory Chart

CANNED FOOD ITEM	QUANTITY	JAR SIZE	DATE CANNED

PRESERVATION METHOD	NOTES

Dehydrated and Freeze Dry Inventory Chart

FOOD ITEM	QUANTITY	CONTAINER USED	DATE PROCESSED	PRESERVING METHOD	NOTES

Frozen Food Chart

FOOD ITEM	QUANTITY	CONTAINER USED	PRESERVING METHOD	NOTES

Year-End Thoughts

Use this page to jot down your thoughts on how the season went overall. Record observances of things that went wrong, but also things that went better than expected. Gathering your thoughts on the season will help prepare you for the next by serving as a reminder of both your successes and your failures.

Things to Improve Next Year

What do you want to do better next growing season? What were your biggest lessons learned? How will you take the successes and failures of this season and translate them into actions to improve your small farm in coming seasons?

Dream Big: My Farm's Plan for Next Year

Here is a dreamer's space. Record your vision for next season's garden. Use it to inspire you when the cycle of your small farm begins anew.

Conclusion

I distinctly recall the moment when gardening fell into place for me. Seated in my garden, I couldn't ignore the disappointment that surrounded me. I questioned myself, wondering why my garden wasn't flourishing abundantly like those of other small farmers I followed on social media. Was it my fault? Did I lack what it took to be one of those efficient, small-scale farmers? After a dramatic self-pity session, I decided enough was enough. I declared that I could indeed cultivate an abundant garden and run a lean, efficient farm. I put an end to the self-pity, understanding that the only obstacle standing between me and success was myself. Now all I needed to do was identify those limiting factors.

I dove into every book I owned on small-scale farming and bought every book by Jean Martin Fortier and Eliot Coleman. I immersed myself in reading, scribbling notes, leaving sticky note reminders, and repeatedly poring over those books for what felt like an entire year. Upon reviewing my notes and highlighted sections, the realization hit me—it was, indeed, *me* hindering the success of my farm.

Naturally, I am a free spirit, guided by what feels right, moved by emotion. The chaotic nature of my life often brings me peace, and I thought I could apply this same free-spirited approach to my gardens. However, the lack of structure resulted in chaos that didn't bring peace; instead, it yielded mediocre crops, and my family still had to rely on other food sources.

Despite the challenges, I recognized that to achieve success like the farmers I admired, I needed to adopt the systems that formed the backbone of their farms. The journey began with a crop plan, a crucial step. No more casually flipping through seed catalogs and choosing the most visually appealing seeds. I decided to invest in seeds of the crops that my family eats, ones I knew would thrive, and for which I had designated space in my crop plan.

Following that, I began using automatic timers for irrigation, creating seasonal checklists, documenting farm expenses, reviewing varieties, establishing chore schedules, time blocking, and managing tasks with utmost efficiency. This transformation wasn't an instant success. Building and establishing these systems, charts, and routines took time. However, once we found our rhythm, the produce literally started overflowing from the beds. We experienced more abundance than I had ever imagined, all while spending less time on the farm. Automation, planning, and maintaining a lean and efficient approach allowed our farm and gardens to work for us, harmonizing with nature.

Inspired by this revelation, I've since made it my mission to encourage and teach as many people as possible about the importance of having a plan and being efficient with your small farm. My hope is that this planner saves you the years of hard work I invested. As you plan your gardens for the year, use this planner as your guide to success. Keep detailed notes and refer to it for friendly reminders. My wish is that one day you'll have your own success story, a moment when everything clicks for you, forever transforming your farming success.

Here's to abundance, friend, because you deserve it!

Jill

"Amidst the petals
of imagination,
dare to cultivate
the wildest dreams,
for in the garden of
possibility, anything
can blossom."
—Anonymous

Conversion Charts

Temperature

To convert Fahrenheit to Celsius and vice versa:

Celsius = Fahrenheit minus 32, multiplied by 5/9
Fahrenheit = Celsius multiplied by 9/5, plus 32

Fahrenheit and Celsius

FAHRENHEIT	CELSIUS
-50	-45.5
-40	-40
-30	-34.5
-20	-29
-10	-23
0	-18
10	-12
20	-7
30	-1
32*	0*
40	4.5
50	10
60	15.5
70	21
80	27
90	32
100	38
110	43
120	49
212†	100†

*Freezing point = 32°F = 0°C
†Boiling point = 212°F = 100°C

Other Conversions

Imperial to Metric Conversions

TO CONVERT FROM	TO	MULTIPLY BY
Inches	Millimeters	25.4
Inches	Centimeters	2.54
Feet	Meters	0.3048
Yards	Meters	0.9144
Square inches	Square centimeters	6.4516
Square feet	Square meters	0.0929
Square yards	Square meters	0.8361
Ounces (liquid)	Milliliters	29.5735
Pints (U.S.)	Liters	0.4731
Pints (U.S.)	Pints (Imperial)	0.8326
Quarts (U.S.)	Liters	0.9463
Quarts (U.S.)	Quarts (Imperial)	0.8326
Gallons (U.S.)	Liters	3.785
Gallons (U.S.)	Gallons (Imperial)	0.8326
Ounces (weight)	Grams	28.3495
Pounds	Kilograms	0.4535

Metric to Imperial Conversions

TO CONVERT FROM	TO	MULTIPLY BY
Millimeters	Inches	0.0393
Centimeters	Inches	0.3937
Meters	Feet	3.2808
Meters	Yards	1.0936
Square centimeters	Square inches	0.1550
Square meters	Square feet	10.7639
Square meters	Square yards	1.1959
Milliliters	Ounces (liquid)	0.0338
Liters	Pints (U.S.)	2.1133
Liters	Pints (Imperial)	1.7597
Liters	Quarts (U.S.)	1.0566
Liters	Quarts (Imperial)	0.8798
Liters	Gallons (U.S.)	0.2641
Liters	Gallons (Imperial)	0.2199
Grams	Ounces (weight)	0.0352
Kilograms	Pounds	2.2046

Length

1 mile = 1760 yards = 5280 feet = 1609 meters = 1.609 kilometers

1 yard = 3 feet = 36 inches = 91.44 centimeters = 0.9144 meters

1 foot = 12 inches = 30.48 centimeters = 0.3048 meters

1 inch = 2.54 centimeters = 25.4 millimeters

Area

1 square mile = 640 acres = 258.9988 hectares

1 hectare = 2.4710 acres = 11,959.9004 square yards = 10,000 square meters = 107,639.1041 square feet

1 acre = 4840 square yards = 4046.8564 square meters = 43,560 square feet = 0.4046 hectares

1 square yard = 9 square feet = 1296 square inches = 0.8361 square meters

1 square foot = 144 square inches = 0.0929 square meters = 0.1111 square yards

An acre is equivalent to the area of a square 208.7103 feet on a side.

A hectare is equivalent to the area of a square 100 meters on a side.

Volume

1 pint = 16 fluid ounces = 2 cups = 473.1764 milliliters = 0.4731 liters = 0.5 quarts = 0.125 gallons

1 quart = 32 fluid ounces = 4 cups = 2 pints = 946.3529 milliliters = 0.9463 liters = 0.25 gallons

1 gallon = 128 fluid ounces = 16 cups = 4 quarts = 8 pints = 3785.4117 milliliters = 3.7854 liters

1 cubic yard = 27 cubic feet = 0.7645 cubic meters = 764.5548 liters = 25,852.6753 fluid ounces

1 cubic foot = 1728 cubic inches = 0.0283 cubic meters = 28,316.8465 cubic centimeters = 28.3168 liters = 957.5064 fluid ounces

1 bushel = 4 peck = 32 quarts (dry) = 64 pints (dry) = 1.2444 cubic feet = 0.3523 cubic meters

1 peck = 8 quarts (dry) = 16 pints (dry) = 0.25 bushels = 537.605 cubic inches = 8809.7675 cubic centimeters

1 imperial bushel = 1.0320 bushels = 8 imperial gallons = 1.2843 cubic feet = 2219.36 cubic inches = 36,368.7943 cubic centimeters = 0.0363 cubic meters

Weight

1 gross ton = 2240 pounds= 1016.0469 kilograms

1 pound = 16 ounces = 453.5923 grams = 0.4535 kilograms

1 ounce = 0.0625 pounds = 28.3495 grams = 0.0283 kilograms

Additional Notes

About the Author

Meet Jill Ragan, the talented author behind *The Tiny but Mighty Farm* and the proud owner of Whispering Willow Farm. With over a decade of hands-on experience in food cultivation, Jill has recently shifted her focus to the beautiful world of cut flowers. Alongside her husband, Nathan, she tends to their lush farm nestled in the heart of central Arkansas, where they cultivate a vibrant online community through platforms like YouTube, Instagram, and their Online Home & Garden Mercantile Store.

What began as a dream and a passion has blossomed into the Ragan family's everyday reality. Jill and Nathan are dedicated to sharing their wealth of knowledge and expertise with others, whether it's through captivating books, immersive workshops, hands-on experiences, or valuable resources. Their hope is to inspire and empower individuals on their own agricultural journeys, fostering a deeper connection to the land and to the joys of sustainable living.

Index